88
Structure and Bonding

Springer

Berlin
Heidelberg
New York
Barcelona
Budapest
Hong Kong
London
Milan
Paris
Santa Clara
Singapore
Tokyo

Metal Sites in Proteins and Models

Iron Centres

Volume Editors:
H. A. O. Hill, P. J. Sadler, A. J. Thomson

With contributions by
S. K. Chapman, M. C. Cox, S. Daff, N. E. Le Brun,
H. Li, G. R. Moore, A. W. Munro, D. P. Nickerson,
A. K. Powell, P. J. Sadler, B.-M. Sjöberg, H. Sun,
A. J. Thomson, A. C. G. Westlake, L.-L. Wong

Springer

In references Structure and Bonding is abbreviated
Struct. Bond. and is cited as a journal.

Springer WWW home page: http://www.springer.de

ISSN 0081-5993
ISBN 3-540-62870-3
Springer-Verlag Berlin Heidelberg New York

CIP Data applied for

Typesetting: Fotosatz-Service Köhler OHG, Würzburg
Cover: Medio V. Leins, Berlin
SPIN: 10552994 66/3020 - 5 4 3 2 1 0 - Printed on acid-free paper

Volume Editors

Editorial Board

Preface

This is the first of 3 special volumes of Structure and Bonding (Volumes 88, 89 and 90) on recent advances in inorganic biochemistry and deals with iron-dependent systems.

There is barely a species which can survive without the use of iron. There are evolutionary chemists, notably G. Wächtershäuser (*Progr. Biophys. Molec. Biol.* 1992, *58*, 85–201), who assert that iron and sulfur provided the energy source to power early life forms before the planet became aerobic. The reduction of protons to hydrogen by sulfide to yield persulfide is catalyzed by Fe(II) ions, the overall process being driven by the insolubility of Fe(II)S_2, ferrous persulfide, more commonly called pyrites or Fool's gold. Although the anaerobic world was dominated by iron–sulfur chemistry, when photosynthsis began to produce dioxygen, biology needed to adapt. Iron became oxidized to a form which is insoluble at neutral pH in the aqueous environment. Hence sequestering methods were needed to scavenge Fe(III). Dioxygen in a reducing world gives rise to highly reactive and potentially toxic reductive products, such as superoxide O_2^-, and hydrogen peroxide. From these species iron catalyzes the formation of even more reactive species, namely hydroxyl radicals. The reactions of O_2^- can be as a low potential oxidant of iron-sulfur clusters, or as a reductant of Fe(III) and hence produce further hydroxyl radicals (according to S. I. Liochev. *Free Rad. Res.* 1996, *25*, 369–384). Biology cannot avoid this chemistry. Instead it utilizes it and controls it. For example the reactions of oxygen and superoxide with Fe–S clusters underlie gene regulation in some bacterial systems and post-transcriptional regulation of the processes of iron uptake and storage in eukaryotes. Metalloproteins which dispose of superoxide and hydrogen peroxide abound in cells to protect, and possibly to sustain, iron-sulfur chemistry. Iron–oxygen chemistry is harnessed by biology to carry out high-activation-energy reactions with organic substrates at ambient temperatures.

Many of these themes are presented in this volume. The transport and storage of Fe(III), to avoid or regulate precipitation, are discussed in the chapters by Moore et al., Sun et al., and Powell. The processes by which Fe(III) is sequestered by transferrin from serum and released within the cytoplasm are discussed in the chapter by Sun, Cox, Li and Sadler. The uptake of Fe(II), its oxidation to Fe(III), and its deposition inside the core of bacterioferritin is outlined by Moore, Le Brun and Thomson. The nature of the inorganic cores of iron-storage proteins and biominerals is described in an elegant article by Powell, which shows the value of appropriate models. The oxidation of Fe(II) in bacterioferritin by O_2 takes place at a dinuclear metal binding site at the centre of an α-helical bundle. This site has remarkable overall similarity to that found in ribonucleotide reductase which utilizes iron–oxygen chemistry to generate a tyrosyl radical and finally to reduce a sugar hydroxyl group before synthesis of DNA is possible. Sjöberg describes the power of crystallographic, site-directed mutagenic, and spectroscopic methods to elucidate the essential functional features of this site.

Only when the functional properties of one site can be engineered into another can we claim to have understanding. This is the message from the chapter by Wong, Westlake and Nickerson in which they describe heme-oxygen chemistry in P-450 enzymes and show how it can be modified to re-direct high energy iron–oxygen chemistry to alternative substrates, some of commercial interest. The diverse and remarkable chemistry of the heme group in a protein lattice is the topic of the chapter by Chapman, Daff and Munro, in which they discuss how the protein controls the reactivity of the heme co-factor either for redox functions or in enzymatic reactions, again with oxygen as the activating substrate.

Overall the volume illustrates the subtle mechanisms which these biological catalysts, stores and regulators employ, and how they can be manipulated and controlled by the chemist to provide new uses and applications.

H. Allen O. Hill, Peter J. Sadler, Andrew J. Thomson

Contents

Contents of Volume 89

Metal Sites in Proteins and Models

Phosphatases, Lewis Acids and Vanadium

Volume Editors: H. A. O. Hill, P. J. Sadler, A. J. Thomson

Contents of Volume 90

Metal Sites in Proteins and Models

Redox Centres

Volume Editors: H. A. O. Hill, P. J. Sadler, A. J. Thomson

Polyiron Oxides, Oxyhydroxides and Hydroxides as Models for Biomineralisation Processes

Annie K. Powell

Centre for Metalloprotein Spectroscopy and Biology, School of Chemical Sciences, University of East Anglia, Norwich NR4 7TJ, UK. *E-mail: a.powell@uea.ac.uk*

Iron biominerals are ubiquitous and fulfil many functions. These are often iron oxide, oxyhydroxide and hydroxide phases, many of which are common in the lithosphere, but some of which, such as the mineral found within the iron storage protein ferritin, are metastable. This article takes the specific example of loaded ferritin as a biomineral and explores the use of model compounds to elucidate the formation, structure and properties of the iron oxyhydroxide mineral contained within the protein. The utility of various compounds is evaluated by considering how closely they match a number of essential criteria defining the composition and properties of loaded ferritin. Those compounds with well-defined structures which can be regarded as "scale models" of the loaded ferritin protein are found to be the most realistic mimics.

Keywords: Iron biominerals; iron oxyhydroxides; nanoparticles; clusters

Structure and Bonding, Vol. 88

1 Introduction

1.1 Iron Biomolecules

Iron is vital to virtually every life-form on Earth, fulfilling a variety of biological functions. Examples include the transport, storage and activation of small gaseous molecules such as oxygen; electron transfer and a range of catalytic processes. Another important group of iron-containing biological molecules include those involved in the uptake, transport and storage of the metal, such as siderophores used by plants and bacteria to achieve iron-uptake, transferrin found in mammals for the transport of the metal and ferritins found in plants, bacteria and animals which can store excess iron. A major influence on the development of this group of molecules involved in iron homeostasis was the evolution of photosynthesis which led to the stabilisation of the +III rather than +II oxidation state of iron as a consequence of the formation of an oxygen-rich atmosphere [1]. This resulted in the predominance of highly insoluble iron(III) oxyhydroxide and oxide phases presenting new challenges for the uptake of the metal previously available terrestrially as much more water-soluble Fe(II) minerals. The problem of maintaining iron supplies in many organisms has been overcome by avoiding loss of the metal as far as possible and recycling it using ferritin as a store. A further problem with the predominance of the +III oxidation state of iron has been the need to avoid significant concentrations of the uncomplexed ion within biological systems since it is capable of catalysing the production of free radicals such as $OH^{\cdot}$ via the Haber-Weiss reaction [2] which can cause cellular damage. Thus we can view ferritins as providing a safe, inert but mobilisable store of Fe(III). Ferritins appear to contain an iron(III) oxyhydroxide mineral of uncertain composition. It is widely held that this biomineral is similar to the "$Fe(OH)_3$" phase initially produced on hydrolysis of iron ions. We shall explore the reasoning behind this in the following sections.If, for any reason, the iron contained within ferritin becomes immobile, it then transforms into what appears to be a better-defined mineral phase, probably goethite, within haemosiderin [3].

1.2 Iron Biominerals

Ferritins can be regarded as members of a final group of iron biomolecules, the biologically produced minerals. These biominerals are predominantly iron oxide and oxyhydroxide phases although other phases such as sulphides like greigite are also observed. All classes of biomineral function for iron oxyhydroxides and oxides are represented [4] and include structural deposits such as the radular teeth of various molluscs made of goethite, α-Fe · O · OH, lepidocrocite, γ-Fe · O · OH, and magnetite, Fe_3O_4; sensors made of magnetic iron oxide particles, usually magnetite, Fe_3O_4, in magnetotactic bacteria, birds, bees, salmon etc.; and metal ion regulation, as demonstrated by iron storage in ferritin probably in the form of ferrihydrite "$Fe(OH)_3$" or $5Fe_2O_3 \cdot 9H_2O$.

1.2.1 *Iron Hydroxide, Oxyhydroxide and Oxide Minerals*

These iron oxyhydroxide and oxide phases all result from the hydrolysis reactions of iron ions [5–7]. Both Fe(II) and Fe(III) can be used as starting material for the production of iron(III) oxyhydroxide and oxide minerals [7]. Hydrolysis of iron(III) leads to the highly insoluble iron(III) "hydroxide" as an initial solid product and will form immediately on addition of a base to an aqueous solution of an iron(III) salt. This material is usually formulated as "$Fe(OH)_3$" and called ferrihydrite, but its exact composition is highly variable and it is highly likely that it is metastable and corresponds to a mixture of oxyhydroxide phases of various degree of hydration. The presence of anions such as chloride, phosphate or sulphate can also influence the nature of the phase formed [5]. Depending on how this material is treated, different well-defined oxyhydroxide and oxide phases can be stabilised. If "$Fe(OH)_3$" is left to age under water, the two well-defined AX_2-type iron(III) oxyhydroxide phases, lepidocrocite, γ-Fe · O · OH, and goethite, α-Fe · O · OH, can be formed. An alternative route to the formation of these is via the ferrous hydroxide, $Fe(OH)_2$, which has the CdI_2 structure, exemplified for hydroxides by the mineral brucite, $Mg(OH)_2$ [7]. The brucite lattice consists of layers containing two strips of hydroxides with metal centres in octahedral holes. The hydrogen-bonding between these layers results in an interdigitation of the hydroxides resulting in a "zipper" effect (Fig. 1) [7]. The transformation of a layered brucite structure, $M(OH)_2$ into a three dimensional M · O · OH structure such as goethite results from removal of half of the hydroxide protons and a relative shift of the metal layers to optimise the sharing of the remaining protons between the layers (Fig. 2) [7]. In this way close-packed layers in goethite are made up of oxygen atoms which bridge three iron centres and share their protons with the oxygen atoms of the next layer. In lepidocrocite, two types of oxygen atoms are discernible. The oxygen atoms within the strips of the oxyhydroxide bridge four iron centres, whilst the two set of oxygens on the outer surfaces of the layers are μ_2 hydroxides. This leads to three different layers and the cubic-close packing of the oxygen atoms (Fig. 3) [7]. Goethite and lepidocrocite are also the important constituents of rust, which is more normally observed on steel as the result of electrolytic reactions on the metal surface. Goethite, the α-phase, can be further

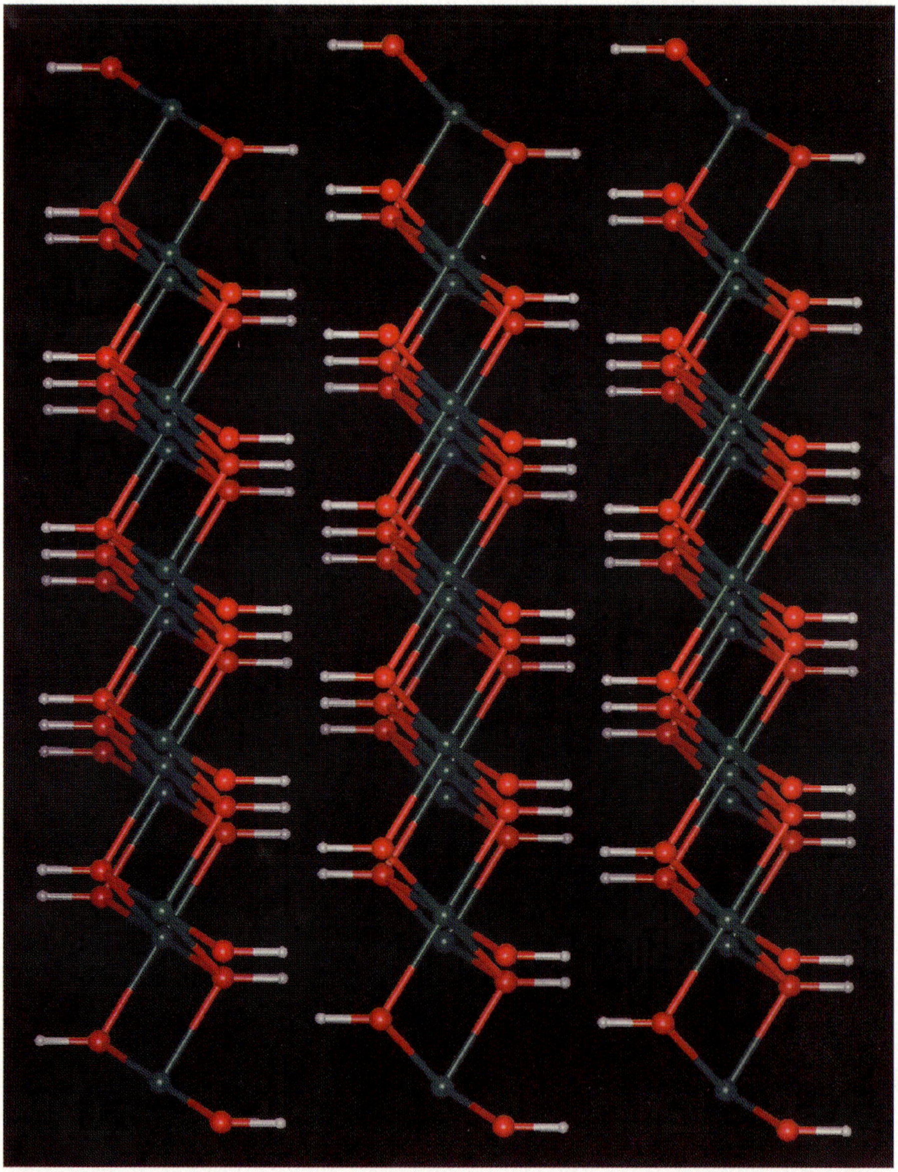

Fig. 1

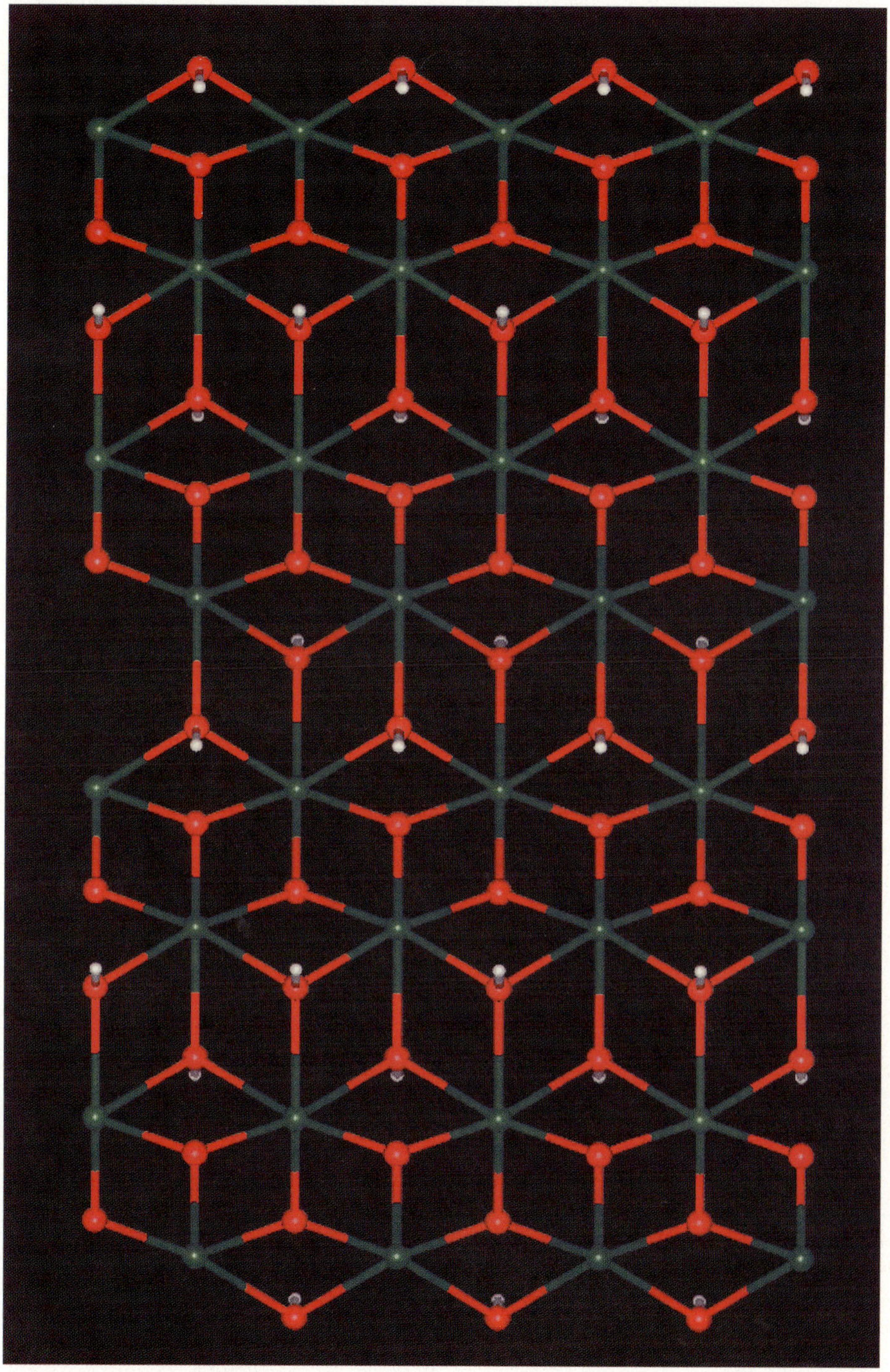

Fig. 2

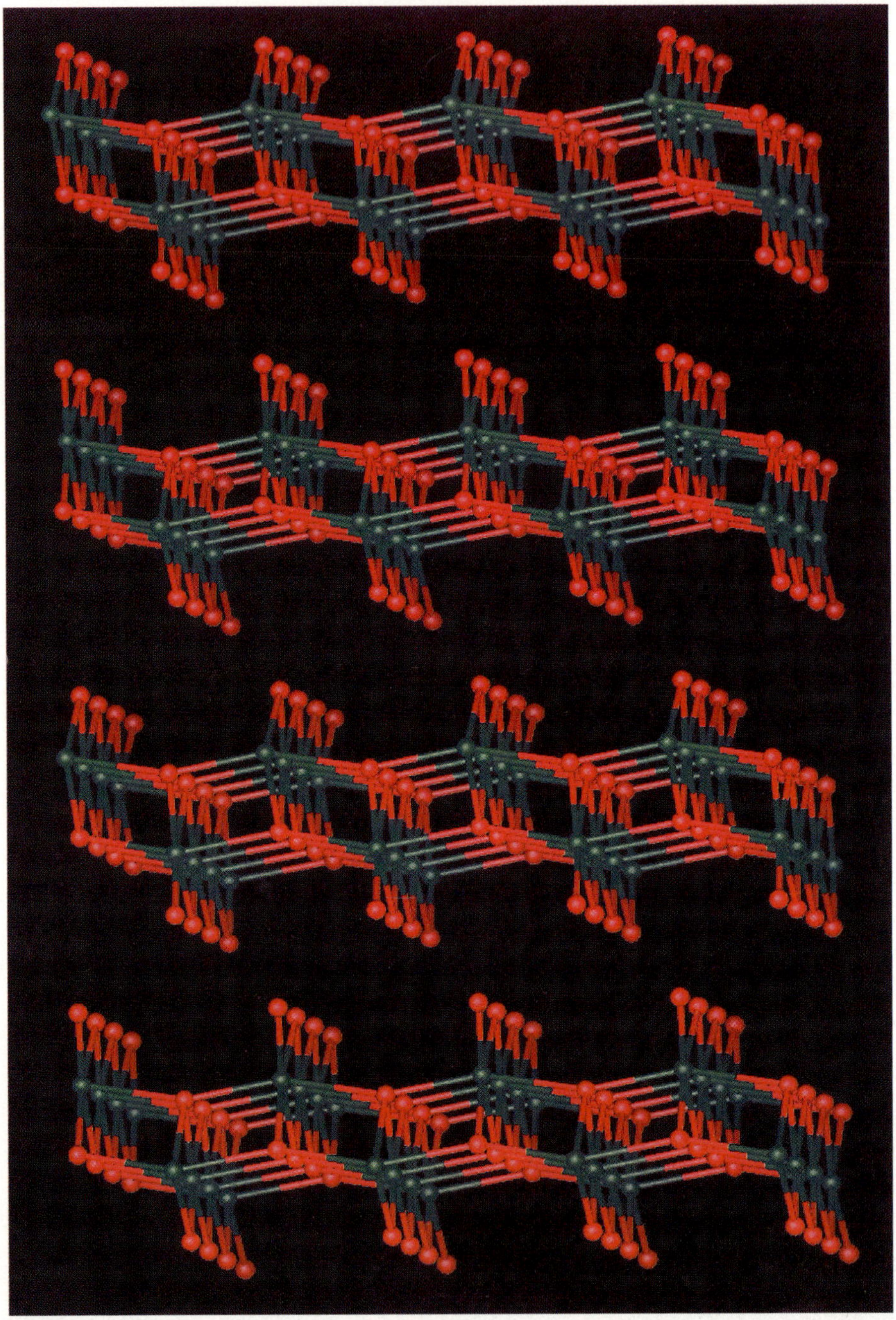

Fig. 3

transformed by heating into the related iron(III) oxide, haematite, α-Fe_2O_3. These α-phase minerals possess hexagonally close packed arrays of oxygen atoms (from the oxides and hydroxides) with iron centres in the octahedral holes. The γ-phase minerals, γ-$Fe \cdot O \cdot OH$ and γ-Fe_2O_3, are based on cubic close packed structures, again with the iron centres in octahedral holes. Maghaemite, γ-Fe_2O_3, is a defect structure, which, under ambient conditions, only forms via the mixed valence magnetite. In fact, the structure of γ-Fe_2O_3 is not very well-defined, but can be understood by considering the oxidation of the iron(II) centres in wüstite, $Fe^{II}O$, which has the sodium chloride structure, but with an Fe:O ratio of less than unity. The charge imbalance is corrected through oxidation of some of the ferrous centres, leading to magnetite (with a cubic inverse spinel structure) and then γ-Fe_2O_3. Thus the relationships of the ferric oxyhydroxide and oxide minerals to ferrous counterparts is clear.

1.2.2
The Relevance of the Hydrolytic Chemistry of Iron(III) in Biomineralisation

Since water is the principal solvent existing in both organisms and the natural environment, the hydrolytic chemistry of iron(III) is of especial relevance. Such media contain or are in contact with ligating groups, e.g. on small molecules such as citrate, humic acids etc. and protein side-chains [8]. At this point we can begin to appreciate that Nature is able to manipulate the hydrolysis reactions of iron(III) to produce iron oxyhydroxides and oxides for specified functions and that various ligating groups can be used to act as templates to direct the structure of the mineral phase. We can summarise the general hydrolysis reaction, where some potential ligand, L^{m-}, is also present, as in Scheme 1 [9]:

$$\{FeL_n\}^{\{mn-3\}} \rightleftharpoons \{Fe_xL_n(O)_y(OH)_z(H_2O)_p\}^{(mn+z+2y-3x)^- \text{ or } (3x-mn-z-2y)^+}$$

$$\downarrow\uparrow$$

$$\text{“}Fe(OH)_3\text{”}$$

Scheme 1

Nature is able to control this hydrolysis so that, for example, the speciation of the iron can correspond to the extreme left hand side of the scheme with the metal in a soluble and transportable form even at physiological pH, e.g. as in a siderophore complex. Alternatively it can be precipitated as a mineralised phase to perform some specific function as a biomineral. It is believed that the templating effect of the ligand is crucial in directing which phase is formed, that is, the influence of the template, L, in the initial species $\{Fe_xL_n(O)_y(OH)_z(H_2O)_p\}$, and this can include metastable minerals which would not normally be observed under ambient synthetic conditions. This ability of Nature to effect chemical reactions not accessible to us under similar conditions in vitro is not surprising when we consider other reactions, such as the fixation of dinitrogen to give ammonia, which *Azotobacter* achieves in high yields at room temperature and pressure but we need to resort to the Haber-Bosch process using conditions of 200 atmospheres and 400 °C with yields of less than 20%. Although the role of the templating ligand matrix seems clear, its mode of action remains something of a mystery. One difficulty is that in condensed phases such as goethite or lepidocrocite it will be difficult to observe the presence of small amounts of the templating matrix. At this point it is hard to discover to what extent the matrix becomes incorporated in the mineral although experiments performed *in vitro* with a variety of growing mineral phases, such as calcite and aragonite (polymorphs of $CaCO_3$), suggest that growth can be inhibited or enhanced by the presence of groups such as carboxylates, which can substitute for the carbonate moieties at the growing surface of the mineral [10]. Additionally, the use of a variety of surfactants to mimic the action of templates in influencing crystal phase and architecture has produced structures displaying some aspects of the intricate frameworks and forms observed in Nature [11, 12].

2 Biomineralisation and Ferritin

2.1 The Special Case of Ferritin as a Biomineral

The situation within ferritin is somewhat different. The iron biomineral now has a substantial amount of organic (matrix) material associated with it and corresponds to an identifiable molecule which we can represent as $\{Fe_xL_n(O)_y(OH)_z(H_2O)_p\}$ with the polypeptide subunits which make up the protein coat equating to L. In fact, the ratio of the template L to mineral is so large that it is now the structure of the biomineral which is hard to detect. Thus, the details of the structure of the surrounding protein coat are long-established but the precise nature of the mineralised core is still open to question. In the following sections these points will be discussed further. By way of introduction, it is worth underlining the fact that gaining an understanding of the nature of the core(s) in ferritin(s) offers our best hope in beginning to understand the way in which the matrix can direct structure. The reason for this is that apoferritin represents a confined reaction vessel in which biomineralisation can take place as well as being very well-characterised structurally [13]. Additionally, many studies have been performed on the iron cores of ferritins so that we have a fairly good idea of our aims when trying to synthesise model compounds [3].

2.2 The Structures of the Ferritin Reaction Vessels

The protein crystal structures of examples of both mammalian, from horse spleen, and bacterial, from, e.g. *E. coli*, ferritins have been determined [14 and references therein]. The structural motifs are essentially identical with 24 subunits forming a spherical protein shell enclosing a cavity of some 7–8 nm in diameter. This cavity is reached by 3-fold and 4-fold channels, as dictated by the highly symmetric nature of the protein structure which crystallises with cubic 432 symmetry in the majority of cases. The channels probably allow for the passage of water to and from the cavity and possibly also iron, although these suggestions are difficult to prove. As yet no structure of a phytoferritin has been reported. However, it seems likely that all these proteins retain the basic motif of 24 subunits forming a reaction cavity [14, 15]. This is an important point, since, *per force*, all experiments on loading ferritins have been performed *in vitro*. It has been demonstrated that several different types of core can be laid down, for example, the mixed valent mineral magnetite (Fe_3O_4) [16] instead of the usual ferric oxyhydroxide, depending on iron source, presence or absence of phosphate, temperature and, indeed, metal with the mixed valent Mn_3O_4 analogue of magnetite also reported [17]. It turns out that, for example, bacterioferritin can be used as a reaction vessel to produce the ferric oxyhydroxide mineral usually found in animal ferritins rather than the iron hydroxyphosphate mineral core observed in native bacterial samples [18–20]. Whilst it seems clear that the function of animal ferritins is to provide a mobilisable iron store, the situation within bacterioferritins is ambiguous, with the large amounts of phosphate associated with the iron core of native samples suggesting that the function of the protein might not be equivalent to that of animal ferritin [21].

2.3 The Nature of the Iron Core of Ferritin

Given that both bacterial and animal ferritins can be made to produce cores of iron oxyhydroxide minerals, we now turn to the difficulties in ascertaining the exact nature of these cores. Although the reaction vessel of apoferritin provides an excellent vehicle for testing theories of biomineralisation processes there is a serious problem in characterising precise structural details of ferritin cores. This is a direct result of the nanoscale proportions of the interior cavity which effectively limit the size of any mineralised core to a maximum of 7–8 nm. Such a nanoscale particle will have properties belying its bounded state. This is demonstrated most clearly in the magnetic behaviour of ferritin cores such as superparamagnetic and quantum tunnelling effects arising from non-zero spin ground states on antiferromagnetically coupled particles. This point is discussed further in Sect. 3.4.5. The magnetic and other spectroscopic properties of ferritin cores have been reviewed in detail by Le Brun, *et al.* [21 and references therein].

There is yet another problem, which is the loading of ferritin cores. It is very difficult to produce uniform loading of ferritin and usually a range of core sizes as determined, for example, by electron microscopy and diffraction is observed, even in samples which have been loaded *in vitro* [19, 22, 23]. The standard estimate of 4500 iron centres for full loading of the 7.8 nm diameter sphere has been calculated on the basis of the core corresponding to the mineral ferrihydrite as is discussed further in Sect. 3.1 below.

The way in which the iron mineral forms is also difficult to elucidate. The various models for the uptake of iron into the cavity have been reviewed recently by Le Brun, *et al.* [21] and Harrison and Arioso [14]. It is known that using iron(II) as a source, as is thought to be the case *in vivo*, an iron(III) oxyhydroxide mineral forms within the protein cavity both in the presence and absence of oxygen. It thus seems likely to this author that it is the hydrolytic chemistry of iron ions which is the key to the formation of this biomineral. Therefore, a study of the controlled hydrolysis of iron ions, using the synthetic strategy suggested by scheme 1, should provide insights into the way in which ferritin cores can be formed as well as the role of the organic matrix in directing structure to provide a mobilisable but unreactive iron store.

Once the iron ions have reached the inner cavity, it is thought that they attach to nucleation sites on the inner surface of the protein shell. The exact identity of these nucleation sites is not known for certain, but it is possible that they are carboxylate residues. The iron mineral will then grow from such sites [24]. From the theories of nucleation and crystallisation it is likely that only one or two crystallites will grow to any size. For crystal growth to occur, the surface energy required to form the particle must be overcome, otherwise the atoms will redissolve. The free energy of crystal nucleus formation can be quantified by considering the balance between a favourable energy term which includes the parameters of particle size, thermodynamic temperature, ionic activity product and stability constant for the mineral, and an unfavourable surface energy term [25]. Once a critical particle size has been achieved, crystal growth can dominate over nucleation and the theory suggests that any new small nuclei which form will redissolve whereas atoms adding to the dominant crystallite(s) will contribute to the continued reduction in the importance of the unfavourable surficial energy term. However, as noted above, it is impossible to tell how fully loaded any given core might be and the overall distribution of crystallite sizes is not known. Thus we are likely to be looking at ferritins with unevenly sized cores which in any case are too small to count as infinite, or condensed structures.

In summary, there is an element of random distribution in operation which is a result of an unequal loading of individual ferritin molecules and maybe a consequence of the way in which ferritin is loaded leading to disorder and inefficient packing of the cavity. Furthermore it is not known whether the iron centres in ferritin form one crystallite or several smaller crystallites. Another difficulty in determining the structure of the mineralised core of ferritin arises from the fact that even if the ferritin is fully loaded with 4500 iron oxyhydroxide units in the core and all of these are in the same crystallite corresponding to a portion of an iron oxyhydroxide mineral, the size limitation of the cavity means that a particle of no more than 8 nm in diameter will form having about 30% of these units lying on the surface of the sphere, with the remaining 70% defining the bulk. Therefore, although the iron oxyhydroxide units in the bulk will be in environments corresponding to those of the iron oxyhydroxide units in the extended mineral lattice, those at the surface will be subject to very marked boundary effects as a consequence of the small size of the particle. This leads to difficulties in defining the structure of the mineral portion which is trapped within the ferritin molecule. Diffraction experiments on the cores will suffer from the fact that the physical interpretation of X-ray diffraction data requires that the assumption can be made that the lattice under study corresponds to a truly infinite array with no boundaries. This is reasonable even for very small single crystals of micrometre dimensions, containing very many more iron oxyhydroxide units than the ferritin core can accommodate, in which over 99.75% of the molecules would be in the bulk with only about 0.25% on the surface and therefore there would be no discernible boundary effects. This is obviously not the case for the ferritin core. The fact that ferritin cores exhibit unusual magnetic properties associated with their nanoscale proportions reinforces this. Ideally, the structural details of the core would best be found using single crystal X-ray diffraction data on fully loaded ferritin as part of a complete protein structure determination. However, because of the problems of achieving uniform iron-loading and the sheer scale of the task, requiring the location of the positions of several thousand iron centres, this route is not currently viable.

We have chosen loaded ferritin to illustrate the way in which a model compound approach can help in elucidating the nature of metal centres in biological systems. The question regarding the identity of the mineralised core in ferritin is an intriguing one and finding the answer is beset with difficulties which arise as a direct result of the nature of ferritin. By proper choice of model, it has been possible to shed some light on both core structure and the way in which the organic template can direct structure and mineral phase. These points will be developed in the following sections.

3 Proposed Models to Elucidate the Structural Features of Loaded Ferritins

As a result of the problems with discovering accurate structural parameters for loaded ferritins, i.e. core plus protein, much effort has been directed towards identifying suitable models. Several model systems which reproduce aspects of the structural features of loaded ferritins have been suggested and are reviewed in the following sections. The salient features which should be reproduced can be depicted as in Fig. 4.

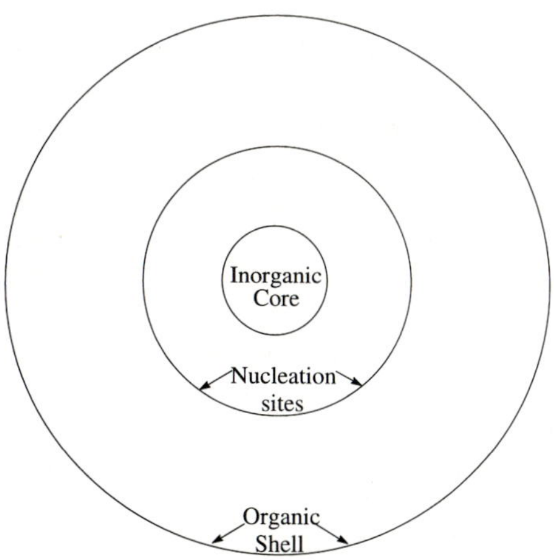

Fig. 4

From this we can see that there are three distinct types of iron centres within loaded ferritins:

(a) There will be the iron atoms located at nucleation sites. These will have coordination spheres containing both organic (from the protein) and inorganic (from the mineral surface) ligands.
(b) There will be iron atoms at the surface of the mineral. These will be connected to the bulk mineral as well as the iron atoms within the protein shell through inorganic oxide/hydroxide linkages and there may be additional linkages to the iron atoms of the shell through dinucleating proteinaceous ligands.
(c) Lastly, there will be the purely inorganic iron atoms which will be in environments corresponding to those of the bulk mineral.

3.1 Iron Oxyhydroxide/Oxide Minerals

Since the ferritin core is composed of an iron(III) oxyhydroxide or oxide mineral, an obvious first step in attempting to model the core would be to examine the properties of naturally occurring minerals. As explained earlier, techniques such as X-ray powder diffraction are of limited use on nanoscale particles such as ferritin cores. However, the experimental fact that it is hard to load more than a few thousand iron atoms into apoferritin also gives an indication of the likely density of the mineral core. If we make the assumption that a fully loaded ferritin can hold 4500 iron centres and we take the preferred mineral model, ferrihydrite, as an example, we can calculate that, based on the formula of $5Fe_2O_3 \cdot 9H_2O$ (corresponding to $96.2\ g \cdot mol^{-1}$ per iron) and a cavity volume of $248\ nm^3$ (internal diameter of 7.8 nm) the density of the mineral would be $2.9\ g \cdot cm^{-3}$, which is on the low side of the range observed for ferrihydrites [26]. The same calculation assuming a formula for the mineral of $Fe \cdot O \cdot OH$ gives a density of $2.7\ g \cdot cm^{-3}$, which is a long way from the experimentally and crystallographically determined values of $4.27\ g \cdot cm^{-3}$, for goethite, α-$Fe \cdot O \cdot OH$, and $3.97\ g \cdot cm^{-3}$, for lepidocrocite γ-$Fe \cdot O \cdot OH$ [27]. Worse yet, the calculation for Fe_2O_3 gives a value of $2.4\ g \cdot cm^{-3}$, which is less than half the value of $5.26\ g \cdot cm^{-3}$, observed for haematite, α-Fe_2O_3 [27]. These numbers suggest that if the cores are truly loaded to capacity with 4500 iron centres, then the mineral type is unlikely to be a pure oxide or even oxyhydroxide such as haematite, goethite or lepidocrocite. We can see that the density of these minerals decreases as we introduce more protons into the lattice, and we can thus expect the ferritin mineral to contain relatively large proportions of these, as ferrihydrite does. The alternative is that the ferritin molecules never achieve full loading. For example, the reverse calculation for the most extreme case of Fe_2O_3 as haematite suggests that a ferritin fully loaded with this mineral would need to accommodate about 9800 iron centres.

3.1.1
Ferrihydrite as a Model for Loaded Ferritin

Whilst the structures of haematite, goethite and lepidocrocite are well-established, that of the favoured mineral model, ferrihydrite, has been the subject of much debate. Over the past thirty years various suggestions regarding the crystal structure of this material have been put forward. As stated before in Sect. 1.2, ferrihydrite is essentially equivalent to "$Fe(OH)_3$", the initial product of the hydrolysis of iron(III) (in the absence of anions such as chloride, sulphate and phosphate, all of which can influence the structure of or become incorporated in the mineral phase). The overall stoichiometry of ferrihydrite is found to correspond to a mixture of $Fe \cdot O \cdot OH$ and Fe_2O_3 plus water molecules, first identified by Towe and Bradley as a possible candidate for the biomineral in ferritin in 1967 and formulated as $Fe_5HO_8 \cdot 4H_2O$ [28]. The most widely accepted formula is $5Fe_2O_3 \cdot 9H_2O$, originally suggested by Chukrov, et al. [29], but the equivalent formula used by Russell of $Fe_2O_3 \cdot 2Fe \cdot O \cdot OH \cdot 2.6H_2O$ underlines the presence of hydroxide as well as oxides [30]. The consensus is that the oxygen atoms form double hexagonally close packed layers with iron centres largely distributed in octahedral holes [31], although recently there has been considerable controversy over whether any of the iron centres occupy tetrahedral holes [26, 32, 33]. The oxygens derive from the oxides, hydroxides and water molecules, the presence of the latter probably accounting for the defect structure with regard to the iron centres. Thus it appears that the material comprises mineral arrays related to the α-phase oxides and oxyhydroxides, haematite and goethite. This is further supported by the fact that ferrihydrite is thermodynamically unstable and can easily be transformed through ageing or the application of relatively mild heating into goethite and haematite [7]. There are some more controversial suggestions that maghaemite can also be formed [26, 32].

A major problem with the structural characterisation of this mineral is the highly variable nature of its composition, with different degrees of hydration observed by different workers. In fact, ferrihydrite is probably best regarded as a mixture of mineral phases and in recognition of this we shall subsequently refer to the material as "ferrihydrites". It has been found that ferrihydrites precipitate with varying particle sizes, ranging from 2.5 nm diameter to 8 nm and larger. Clearly, these dimensions are within the range required for ferritin core models, but we are now faced with exactly the same problem as stated earlier in Sect. 2.3: the particles are too small for boundary effects to be ignored. Indeed, a significant stumbling block in attempting to clarify the structure of the mineral from X-ray powder diffraction experiments is that their small dimensions leads to severe line-broadening; a factor pointed out by mineralogists for many years [34], thereby exacerbating the difficulties in defining the structure(s). Thus the X-ray powder patterns measured for ferritin cores and for ferrihydrites provide only a crude fingerprint and very little in the way of concrete structural information.

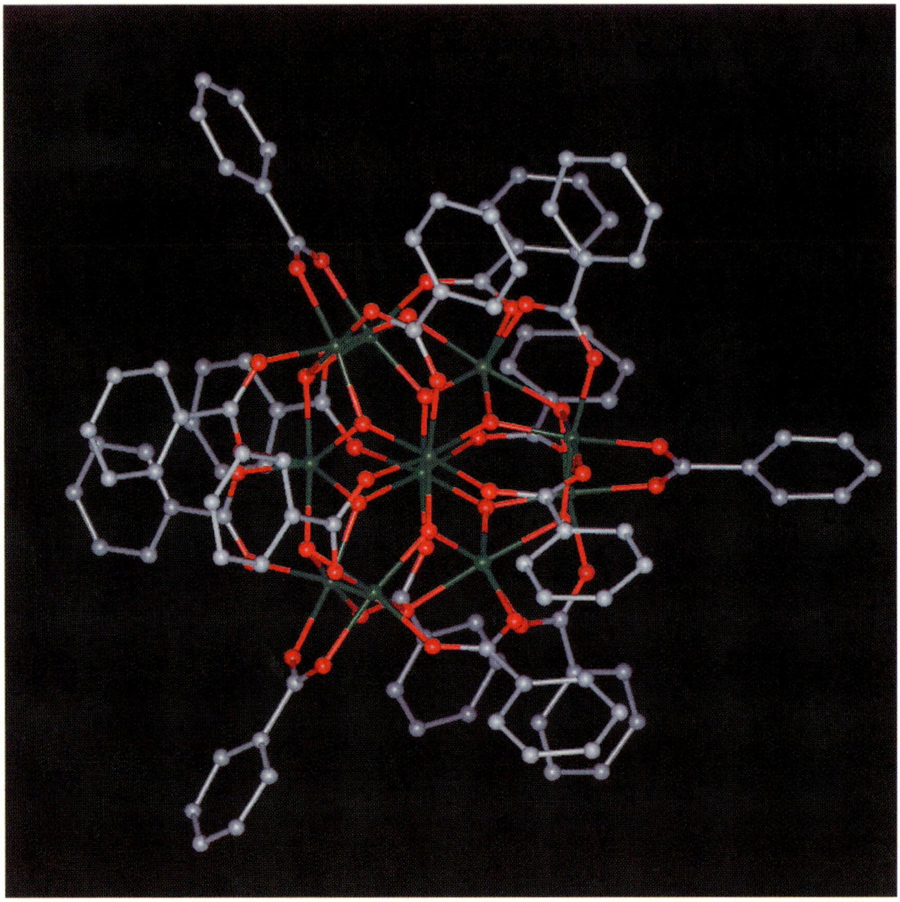

Fig. 5

compound can be viewed as possessing a "pseudo" close-packed mineral core if we imagine replacing the bridging methoxides with hydroxides. This approach can also be applied to the decanuclear iron(III) compound which can be prepared in methanol with the dione ligand dibenzolylmethane (Hdbm) present, and has the formula $[Fe_{10}O_4(OCH_3)_{16}(dbm)_6]$, Fe_{10}(can), (Fig. 8) [43]. Two other sorts of decanuclear iron(III) aggregates have been reported. When small carboxylato ligands, such as acetate or chloroacetate, are present in the synthesis and the solvent is methanol, cyclic products, dubbed as "ferric wheels" of general formula $[Fe_{10}(OMe)_{10}(O_2CX_3)]$, Fe_{10}(wheel) (Fig. 9), can be produced [44, 52] even in the presence of potential ligands such as 6-chloro-2-pyridonate, chp^-, [44]. If conditions are changed, acetonitrile is the solvent and the bulkier benzoate is used as carboxylate a different structure can be isolated from the chp system. This also contains ten iron(III) centres and is related to the Fe_{11} structure. The cluster also incorporates two sodium ions; $[Fe_{10}Na_2(O)_6$

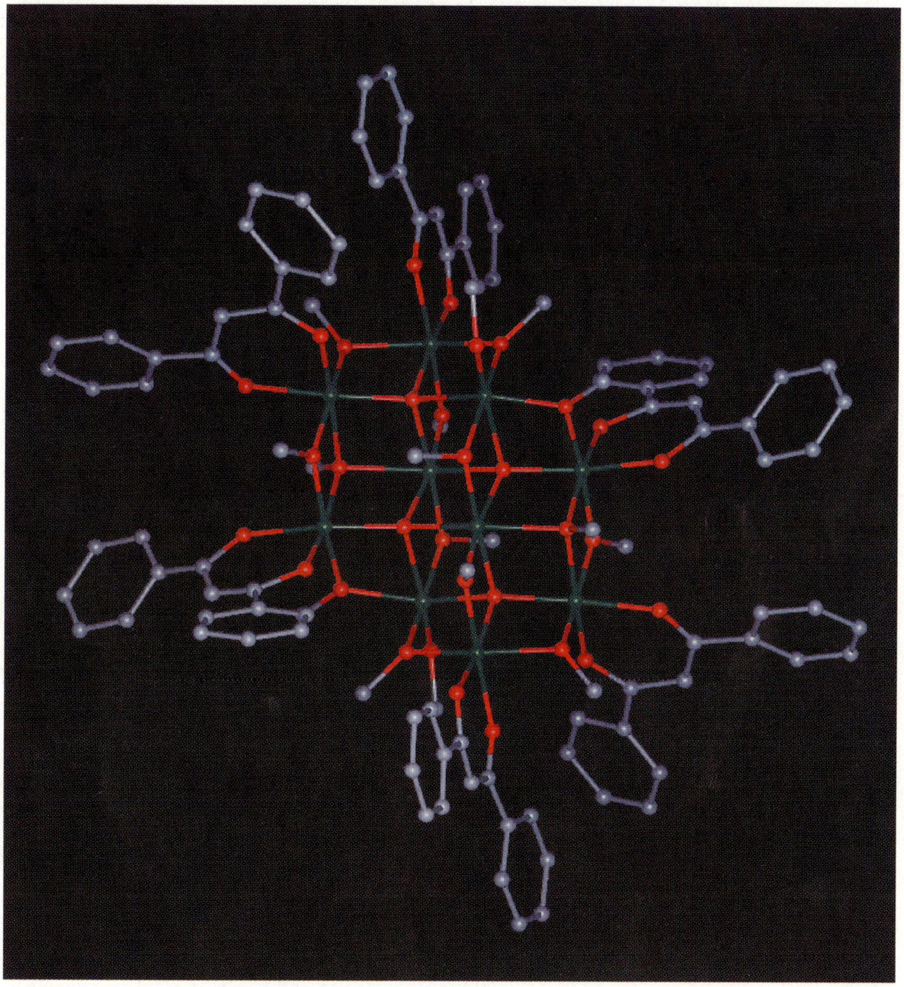

Fig. 8

wholly satisfactory. Similar structural motifs of six-membered Fe_3O_3 and four-membered Fe_2O_2 rings can be seen for these Fe_{17}(repw) and Fe_{10}(repw) aggregates.

3.3.2
Hydrolytic Polyiron Cluster Aggregates from Aqueous Media

The synthetic approach utilising the modified hydrolysis of iron(III) uses chelating ligands to trap mineralised iron species corresponding to

$$\{Fe_xL_n(O)_y(OH)_z(H_2O)_p\}^{(mn+z+2y-3x)-\text{ or }(3x-mn-z-2y)+}.$$

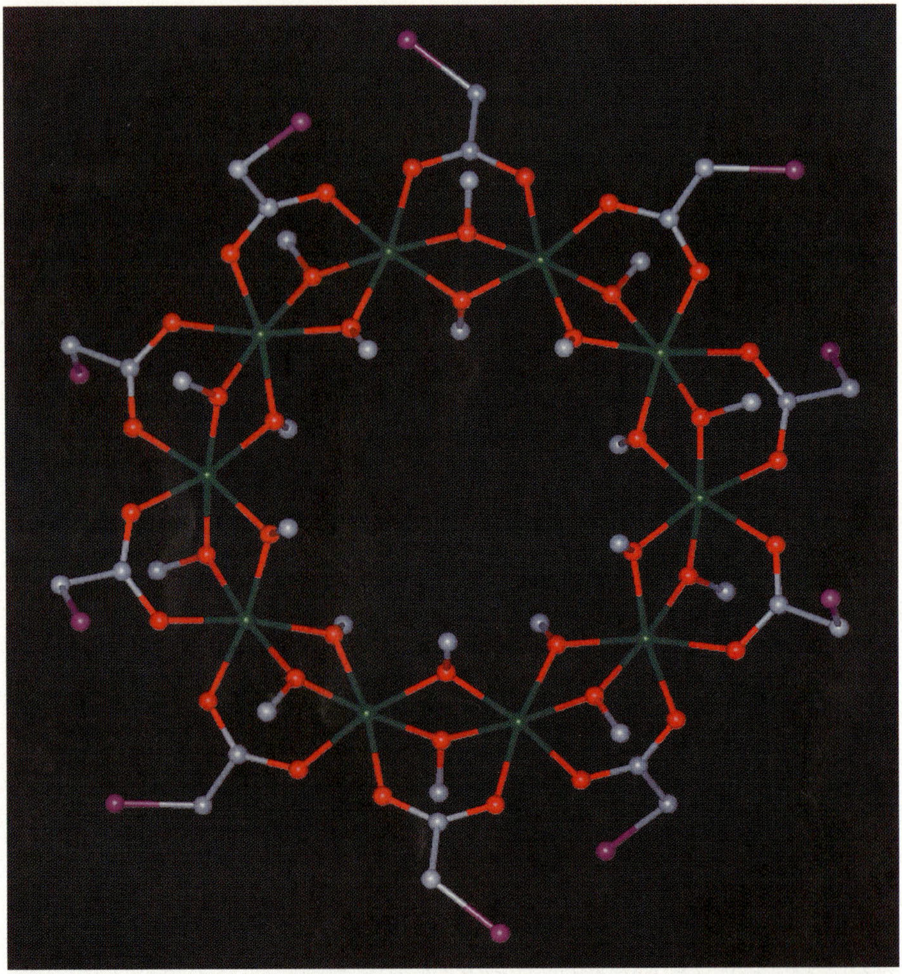

Fig. 9

This approach is fundamentally different from that adopted by groups using non-aqueous routes because it utilises the natural hydrolysis of iron(III) to provide the cluster species. In this system the water is the most concentrated species present whereas in the non-aqueous syntheses it will be the least concentrated. Thus in a typical syntheses the concentration of the iron will be between 0.1 and 0.2 M whereas the concentration of the water is 55 M. The compound isolated from aqueous solutions of iron(III), the ligand 1,4,7-triazacyclononane, tacn, in the presence of counterions such as bromide or iodide, $[Fe_8(O)_2(OH)_{12}(tacn)_6]^{8+}$, Fe_8, was an important early example of this synthetic strategy (Fig. 12) [53]. Smaller aggregates produced from hydrolysed iron(III) solutions with alcohols and diols as ligands have also been reported [54], but are not considered further here.

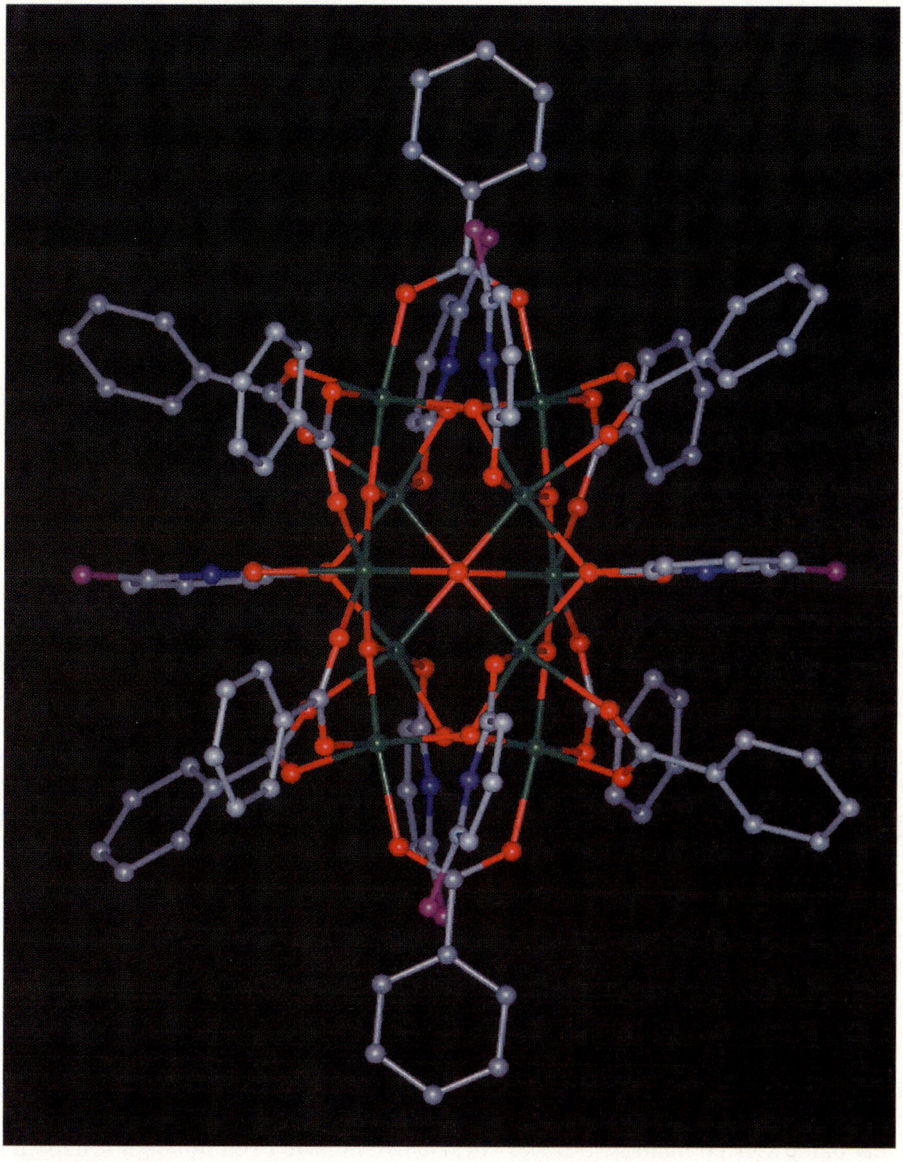

Fig. 10

Using the ligand hydroxyethyl(iminodiacetate), "heidi^{3-}", much larger cluster aggregates containing 17 and 19 iron centres have been isolated, $[Fe_{17}(heidi)_8(O)_4(OH)_{16}(H_2O)_{12}]^{3+}$ and $[Fe_{19}(heidi)_{10}(O)_6(OH)_{14}(H_2O)_{12}]^+$, (Figs 13, 14) [55, 56]. These two aggregates crystallise together and have been shown to possess identifiable mineral cores, which can be thought of as corresponding to an iron(III) hydroxide lattice of the brucite $(Mg(OH)_2)$ type, $\{Fe(OH)_2\}_n^{n+}$. They are

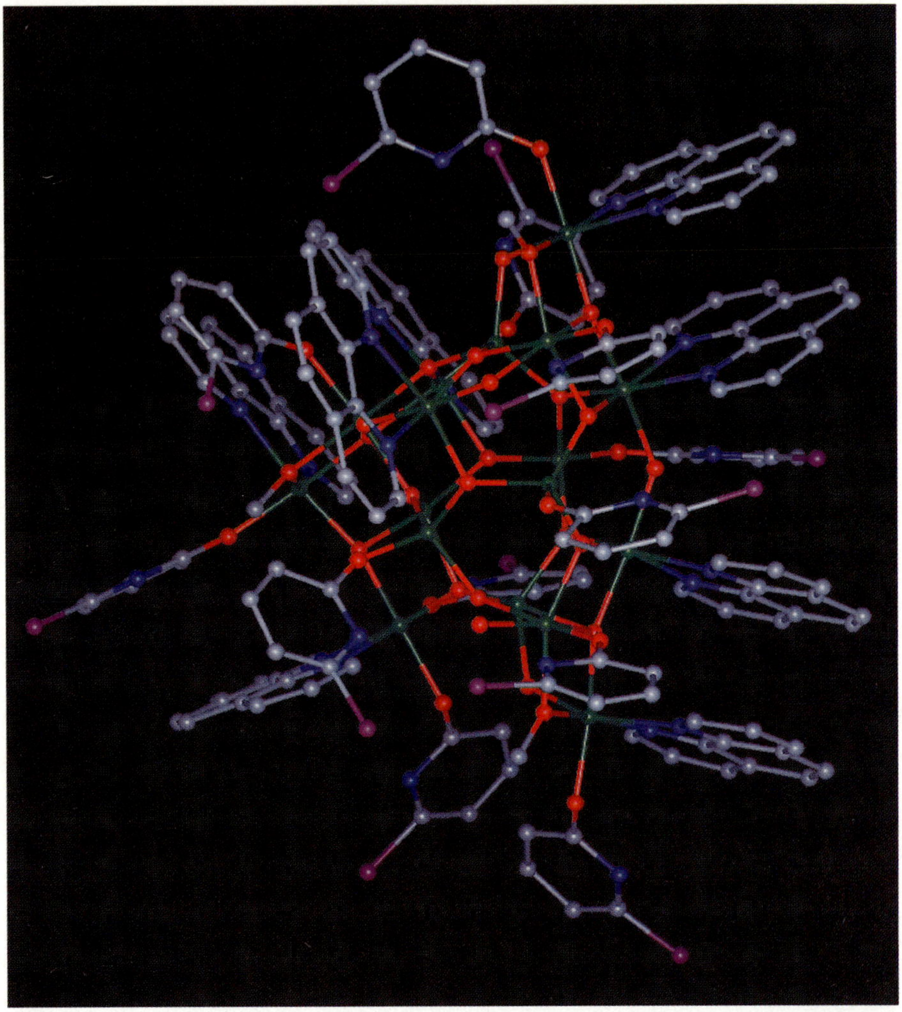

Fig. 11

the best characterised of a range of aggregates which can be synthesised in this way using related ligands [56, 57]. These clusters have been dubbed "crusts" in recognition of the difficulty in nomenclature and the fact that they represent *c*aptured *rust*, and will be identified here as Fe_{17}(crust) and Fe_{19}(crust).

3.3.3
The Effect of the Organic Template

On examining the structures of the various cluster aggregates it can be seen that the nature of the peripheral organic ligands can have a profound effect on the nature of the iron oxyhydroxide aggregate. Amongst the important factors are

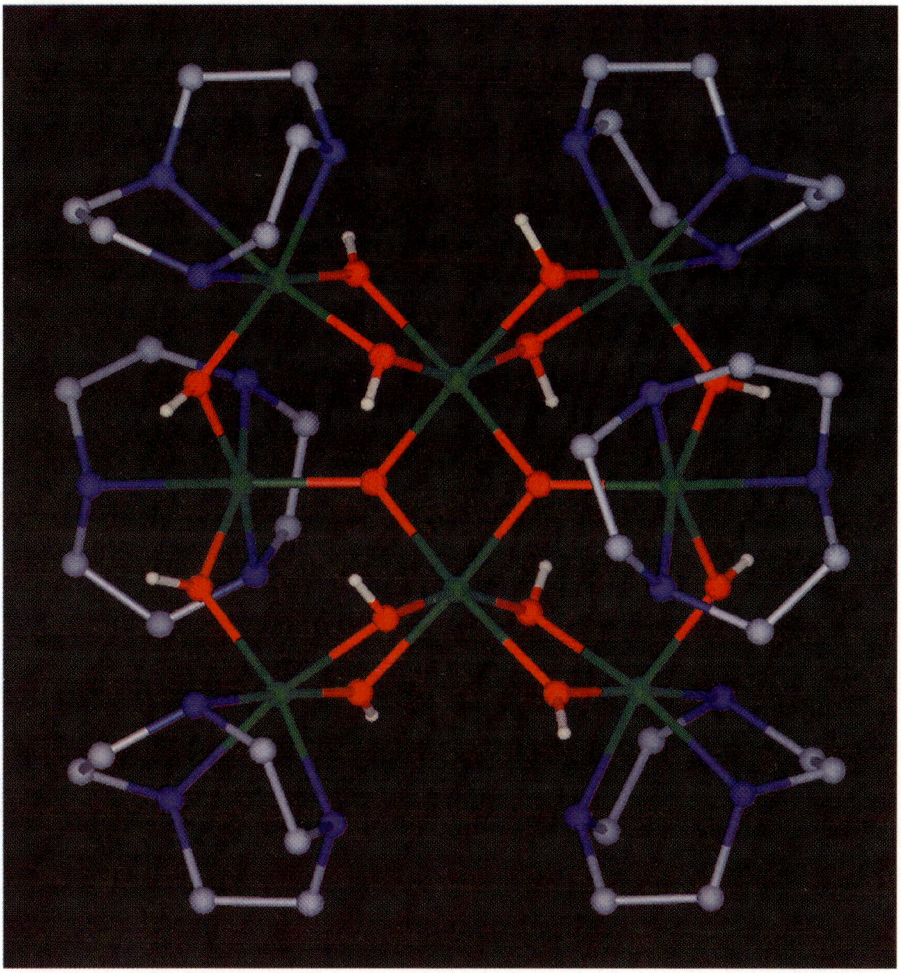

Fig. 12

the charge and denticity of the ligand as well as the local pH in the reaction solution. Naturally, this last parameter is hard to define in non-aqueous systems, but is likely to influence whether the cluster contains predominantly oxide or hydroxide bridges. These effects are easiest to appreciate in the aggregates containing recognisable portions of mineral structures. The effect of a tetradentate ligand such as $heidi^{3-}$ in Fe_{17}(crust) and Fe_{19}(crust) is to block four coordination sites on the iron octahedron, favouring a layered type of structure, whereas the Fe_{12} and Fe_{10}(can) structures both contain portions of three dimensional arrays probably as a consequence of the bidentate peripheral ligands (acetate or dbm^-) allowing for the introduction of further close packed oxygen layers. This templating effect is obviously an important parameter to investigate in future model studies.

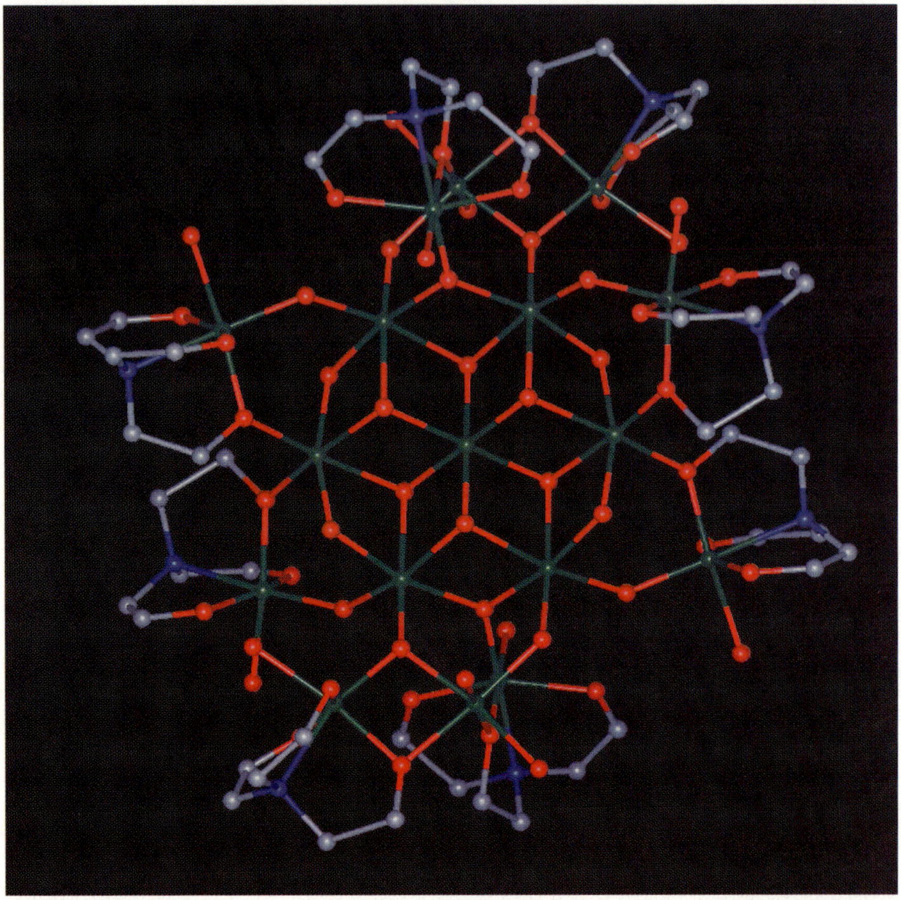

Fig. 13

3.3.4
Properties to Model

In order to identify cluster aggregates which could provide realistic scale models for loaded ferritins we consider first the properties a good model would reproduce.

(a) The system should contain a mineralised but bounded, nanoscale sized core.
(b) The core should be enclosed in an organic shell containing C, H, N and O atoms and some of the O atoms should be carboxylates.
(c) The approximate ratios of Fe:C:N:H:O and the proportions of Fe and inorganic (mineral) core in relation to the total system should be reproduced.
(d) The different sorts of iron environment (as outlined above) should be reproduced, preferably in the correct proportions.
(e) The model compound should possess nanoparticle properties such as anomalous magnetic behaviour.

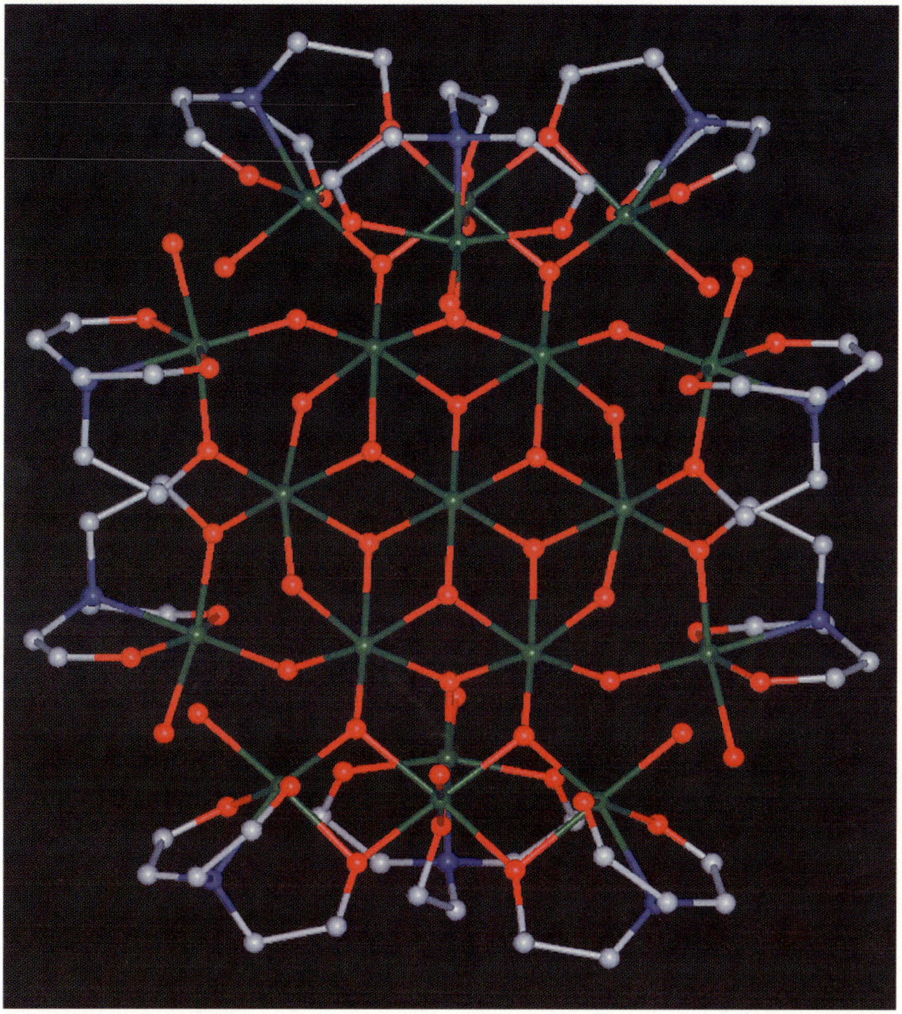

Fig. 14

If all five points are successfully reproduced the cluster molecule would represent a miniature structural model for ferritin. To be of any real use, the model would also need to be well-defined and preferably structurally characterised by single crystal X-ray diffraction.

(f) A final property to model would be that the system could reversibly assimilate and release iron. This would then produce a functional as well as structural model.

As has been indicated above, extended mineral structures are inappropriate as models for loaded ferritins as they fail to reproduce the different iron environments found in bounded structures and naturally cannot be expected to model any features of the protein shell. In fact, it is clear from these considerations that condensed mineral phases such as goethite, lepidocrocite, haematite and so on fail to provide realistic models on all of these counts. Worst still, some of these phases, such as maghaemite, are not well-defined from the structural point of view. Although it is possible to produce polydisperse nanoparticles of iron hydroxide, often in the form of ferrihydrites, these do not fulfil requirements (b) or (c) and only partially fulfil the requirements of points (a) and (d), possessing the correct dimensions and a number of different iron environments, but failing to provide an organic coat. The Saltman-Spiro balls, phospholipid systems and iron-delivery drugs do possess organic coats, but these are not well-defined and also fail to provide any nitrogen donors. In addition, the nature of the mineral phase(s) in such colloidal dispersions will be hard to ascertain since we now arrive back at the problem of structure determination on mineral portions exhibiting large boundary effects.

Clearly we need to concentrate on bounded particles, but these must also be of a size which allows for their complete characterisation. Thus the use of "scale models" in the form of iron cluster aggregates can be of assistance, since compounds accessible to single crystal structure determinations can be synthesised and, in some cases, their structural and physical properties used to shed light on the chemistry and physics of loaded ferritins.

3.4
An Assessment of the Accuracy of the Various Cluster Aggregate Models

The suitability of the various cluster aggregate models will be considered according to how well they fit the criteria (a) to (e) above.

3.4.1
Criterion (a): The Mineral Core

Out of the possible cluster models, we can identify Fe_{10}(can) [43], Fe_{12} [48], Fe_{17}(crust) [55] and Fe_{19}(crust) [55] as containing cores which relate to minerals. We note that Fe_8 [53] can also be regarded as possessing features of close-packed mineralised structures if we take the ligand N atoms into consideration. The Fe_{10}(can) and Fe_{12} structures actually possess methoxide rather than hydroxide core ligands, but the packing of the core oxygens is clear. Fe_{10}(can) displays what can be regarded as three layers of close packed oxygen atoms with iron centres in the octahedral holes. This is an AX_2 structure with parallels to the cubic close packed lepidocrocite lattice. Fe_{12} also displays a cubic close packed structure as a portion of the sodium chloride structure, with oxygens in cubic close packed layers and metals in octahedral holes as a 1:1, AX, lattice. The two crust molecules both possess the same portion of a layer of what is best described as the brucite, $Mg(OH)_2$, lattice at the centre of a "doughnut" consisting of the organic coat, iron nucleation sites and the iron centres which connect on to the surface of the mineral. It should be recognised that without a further layer of oxygen atoms it is not actually possible to identify whether the packing is truly hexagonal or cubic close packed, but the sheet-like array of iron/hydroxide units matches the appearance of one layer of the brucite structure so closely that this is the preferred description.

The Fe_{11} [50], Fe_{10}(repw) [44], Fe_{17}(sjl) [46] and Fe_{17}(repw) [45] structures all contain loosely packed Fe/O cores which do not appear to correspond to known mineral types, but may provide useful models nonetheless, since the proportions of iron and oxygen are quite high. The cyclic nature of the Fe_{10}(wheel) [44, 47] aggregates makes their structures too open to provide good models for mineralised cores.

3.4.2
Criterion (b): The Organic Shell

All of the models possess an organic shell, but for Fe_{10}(wheel) [44, 47], Fe_{10}(can) [43], Fe_{11} [50], Fe_{12} [48], and Fe_{17}(sjl) [46] there is no nitrogen present in the shell and for Fe_8 [53] there is no oxygen. Although Fe_{17}(repw) [45] has oxygen donors in its shell these are not from carboxylates. Only Fe_{10}(repw) [44] and Fe_{17}(crust) [55] and Fe_{19}(crust) [55] possess shells with all of the required elements including carboxylates. At this point we will discard the Fe_{10}(wheel) clusters from the argument on the grounds that they fail to meet the first two criteria.

3.4.3
Criterion (c): The Relative Composition

The easiest property to compare is the average composition of the constituent elements of a given model. These are detailed for loaded ferritin and the selection of cluster aggregates which might provide realistic models in Table 1. There are several useful guidelines to the appropriateness of any given model, including the proportion of the total which represents the inorganic core and the proportion of iron. The relative ratio of iron to core is also useful in assessing the suitability of a model as it gives a rough indication of whether the constitution of the inorganic part of the system mimics that of loaded ferritin even if the overall proportion of core does not. In terms of %iron and %core, the Fe_{17}(crust) and Fe_{19}(crust) come closest to the values for loaded ferritin. Although the proportions of iron for Fe_{17}(sjl) and for Fe_{10}(can) with hydroxides in place of the methoxides are in quite good agreement with that for ferritin, the proportion of core is rather low. If we consider the Fe_{10}(can) model with methoxides there is an improvement in the core proportion, as might be expected, but the %Fe is now too low. Turning to the ratio of iron to core, we can see that Fe_8, Fe_{10}(can) with OH, Fe_{17}(crust) and Fe_{19}(crust) are all reasonably close to the value calculated for ferritin. The nearest values from the model compounds are that of Fe_{17}(crust) and the value calculated for Fe_{10}(can) assuming the methoxides are hydroxides. These models contain cores which can be related to AX_2 mineral structures. The real Fe_{10}(can) cluster, with methoxides, has an iron to core ratio lower than, but still close to, that found in ferritin, which suggests that the additional layer of close-packed oxygens in this model is a feature worth trying

Table 1. A comparison of the composition of loaded ferritin and cluster aggregate "scale" models

Sample with formula. The models are given in order of decreasing nucelarity.	Total mass/ $g.mol^{-1}$ (shell + core)[a]	%Fe	%core[b]	ratio of Fe:core	Formula in terms of $\{Fe_x(O)_y(OH)_z(H_2O)_pL_n\}$ Formula scaled to Fe = 4500 x	y	z	p	n with ligand composition total of scaled y + z + p
Loaded ferritin	913000[c]	28	47	1:1.68	4500	5625	2250	2925	24 polypeptide subunits y + z + p = 10800
Fe_{19}(crust) $[Fe_{19}(heidi)_{10}O_6(OH)_{14}(H_2O)_{12}]^+$	3353	32	48	1:1.50	19 4500	6 1421	14 3316	12 2842	10 $C_6H_8NO_5^{3-}$ ligands y + z + p = 8470
Fe_{17}(crust) $[Fe_{17}(heidi)_8O_4(OH)_{14}(H_2O)_{12}]^{3+}$	2895	33	52	1:1.56	17 4500	4 1059	16 4235	12 3176	8 $C_6H_8NO_5^{3-}$ ligands y + z + p = 7579
Fe_{17}(repw) $[Fe_{17}O_{15}(OH)_6(chp)_{12}(phen)_8(OMe)_3]$	4497	21	30	1:1.43	17 4500	15 3970	6 1588	-	12 C_6H_3ClNO and 8 $C_{12}H_8N_5$ ligands y + z + p = 5585
Fe_{17}(repw) with OH[e] $[Fe_{17}O_{15}(OH)_9(chp)_{12}(phen)_8]$	4455	21	30	1:1.43	17 4500	15 3970	9 2382	-	12 C_6H_3ClNO and 8 $C_{12}H_8N_2$ ligands y + z + p = 6352
Fe_{17}(sjl) $[Fe_{17}O_{10}(OH)_{10}(O_2CPh)_{20}]$	3702	26	35	1:1.35	17 4500	10 2647	10 2647	-	20 $C_7H_5O_2^-$ ligands y + z + p = 5294
Fe_{12} $[Fe_{12}O_2(OCH_3)_{18}(O_2CCH_3)_6(CH_3OH)_5]$	1770[d]	38	71	1:2.03	12 4500	2 750	-	-	6 $C_2H_3O_2^-$ and 5 C_2H_4O ligands y + z + p = 750
Fe_{12} with OH[e] $[Fe_{12}O_2(OH)_{18}(O_2CCH_3)_6(CH_3OH)_5]$	1518	44	66	1:1.50	12 4500	2 750	18 6750	-	6 $C_2H_3O_2^-$ and 5 C_2H_4O ligands y + z + p = 7500
Fe_{11} $[Fe_{11}O_2(OH)_6(O_2CPh)_{15}]$	2629	23	31	1:1.35	11 4500	6 2454	6 2454	-	15 $C_7H_5O_2^-$ ligands y + z + p = 4908
Fe_{10}(can) $[Fe_{10}O_4(OCH_3)_{16}(dbm)_6]$	2459	23	45	1:1.96	10 4500	4 1800	- -	-	6 $C_{15}H_{12}O_2^-$ ligands y + z + p = 1800
Fe_{10}(can) with OH[e] $[Fe_{10}O_4(OH)_{16}(dbm)_6]$	2235	25	40	1:1.60	10 4500	4 1800	16 7200	-	6 $C_{15}H_{12}O_2^-$ ligands y + z + p = 9000

Table 1 (continued)

Sample with formula. The models are given in order of decreasing nucelarity.	Total mass/ $g.mol^{-1}$ (shell + core)[a]	%Fe	%core[b]	ratio of Fe:core	Formula in terms of $\{Fe_x(O)_y(OH)_z(H2O)pL_n\}$ Formula scaled to Fe = 4500				
					x	y	z	p	n with ligand composition total of scaled y + z + p
Fe_{10}(repw)[f] $[Fe_{10}O_6(OH)_4(H_2O)_2(O_2CPh)_{10}(chp)_6]^{2-}$	2813	27	37	1:1.37	10 4500	6 2700	4 1800	2 900	10 $C_7H_5O_2^-$ and 6 C_6H_3ClNO ligands y + z + p = 5400
Fe_8 $[Fe_8(O)_2(OH)_{12}(tacn)_6]^{8+}$	809	55	84	1:1.53	8 4500	2 1125	12 6750	-	12 $C_6H_{12}N_3$ ligands y + z + p = 7875

[a] The mass of the total cluster aggregate only, i.e. excluding counterions and solvent.
[b] Where core refers to the "inorganic derived" atoms in the cluster (but see note e).
[c] Assuming the core contains 4500 iron centres in the form of Fe_2O_3.2Fe.O.OH.2.6H_2O.
[d] Value has been rounded up, ignoring disorder on coordinated methanol molecules.
[e] Replacing methoxides with hydroxides to give an inorganic core.
[f] Ignoring two sodium ions and their associated acetone molecules.

to reproduce in larger clusters. For Fe_8 we see that the ratio of iron to core is quite well reproduced, but the overall proportion of iron and core are too high. This is partly a result of the ligand coat being rather small, and the fact that the organic portion of this cluster does not contain any oxygen atoms means that Fe_8 is not a particularly good model for loaded ferritin. It is also probably on the lower limit of what we might regard as a reasonable size for such a model. For the other Fe_{17} models we find that these contain small amounts of core and that the proportion of iron in the whole core is quite large leading to a much lower iron to core ratio. A similar situation is observed for Fe_{10}(repw) and Fe_{11} so that, overall, these possess too much ligand coat and not enough inner core. This leaves us with Fe_{12} which has an iron to core ratio of about 1:2 indicative of a large amount of core. Even if we replace the methoxides with hydroxides, we still have values for %Fe and %core much higher than those calculated for ferritin, although there is a big improvement in the iron to core ratio. This is likely to be the consequence of the mineral type found in the core of this cluster. Since this corresponds to an AX mineral type with the NaCl structure, we expect a correspondingly greater density of iron atoms per oxygen. This is clearly seen in the crystal structure of this molecule, where the central core has μ_6-O bridges, as we would expect in the NaCl structure (e.g. wüstite, FeO), rather than the μ_4-O bridges which would be observed in AX_2 or A_2X_3 mineral types such as goethite and haematite [58].

If these compounds are formulated according to $\{Fe_x(O)_y(OH)_z(H_2O)_pL_n\}$ then we see that none of the models appear to reproduce the relative ratios of oxide, hydroxide and water molecules particularly well. However, a note of caution should be sounded here since discovering the real nature of the mineral core in ferritin is actually the purpose of these model building attempts. Thus, we have put forward one of the more acceptable formulations for a ferrihydrite phase as a guide only. It is also instructive to compare the total number of oxides, hydroxides and water molecules. Overall, the models with AX_2 type cores, Fe_{17}(crust), Fe_{19}(crust) and the Fe_{10}(can) model with hydroxides replacing the methoxides, come closest to the calculation for ferritin.

From these considerations it is clear that the clusters which contain identifiable close-packed AX_2 mineral structures model the situation for loaded ferritin better than either those with AX mineral types, or more amorphous core structures. Indeed, those with the amorphous core structures might prove useful as models for partially loaded ferritin. The fact that the clusters with AX_2 mineral structures are the better models is not surprising given that the ferritin core is likely to contain an iron oxyhydroxide mineral with a defect hexagonally close-packed structure AX_2 lattice.

3.4.4
Criterion (d): The Different Iron Environments

The different types of iron environment we can envisage in loaded ferritin have been described at the beginning of Sect. 3.

It is stressed here again that the fundamental difference between a bounded biomineral system such as loaded ferritin and an infinite condensed biomineral phase such as the iron oxyhydroxide deposits forming structures such as the radular teeth of molluscs lies in the relative proportions of these iron environments. For a bounded phase such as loaded ferritin, the proportion of iron atoms associated with the organic matrix plus those at the surface form a significant percentage of the bulk material whereas the opposite is true in an extended biomineral system.

EXAFS spectroscopy has been used to explore the iron environments in loaded ferritins and model compounds [39, 59–63]. These environments can be best explored using a combination of well-defined structures of model compounds and EXAFS spectroscopy. In the most recent study it was found that the Fe_{19}(crust) and Fe_{17}(crust) cluster system which has been structurally characterised by single crystal X-ray diffraction, could be used to calibrate the parameters used in the interpretation of EXAFS data [59]. In many earlier iron-edge EXAFS studies on loaded ferritin the focus had mainly been on the Fe-O interactions but later work suggested that the Fe-Fe interactions might provide a better guide [61]. When the model used was the $Fe_{17/19}$(crust) system, this was also found to be the case [59]. In spite of the accurate structural parameters available for $Fe_{17/19}$(crust) it was not possible to identify the individual Fe-O and Fe-N interactions in the EXAFS without introducing an unacceptable overparameterisation. The main reason for this was the range of different Fe-O/N bond lengths observed in the $Fe_{17/19}$(crust) which are a natural consequence of the differing iron environments throughout the clusters. Since it is this aspect which is uppermost as a modelling requirement, it was decided that the iron-iron interactions should be used as a guide. Thus the EXAFS spectra recorded for the $Fe_{17/19}$(crust) were given an optimum fit using the known structural parameters. This was something of a departure from many of the previous treatments where no structural information had been utilised in this way. The overall result was that the underestimation of the Fe-O/N and Fe-Fe interactions resulting from a standard EXAFS interpretation was revised. The approach was tested by using the same parameters for the well-defined minerals goethite and lepidocrocite where it was also found that an improved correlation between the EXAFS and crystallography was achieved. These minerals were included in the study as possible models for the ferritin core and the results were also compared with previously reported data on Fe_{11}. Comparative EXAFS studies on this material and Horse Spleen Ferritin (HSF) have been performed. The initial conclusion had been that Fe_{11} showed too many iron-iron contacts to model the ferritin EXAFS completely satisfactorily [60], but the more recent EXAFS studies on HSF suggest the agreement is better than originally believed [59, 61]. Overall the analysis of the various models for loaded ferritin and the comparison with HSF [61] revealed that the best fit for the HSF EXAFS was obtained using the $Fe_{17/19}$(crust) system, although the Fe_{11} system also provided a reasonable fit. Infinite lattices such as goethite and lepidocrocite were wholly unsatisfactory as models, underlining the requirement for model systems to be bounded particles displaying several different iron environments rather than extended arrays with perhaps just one or two highly symmetrical iron sites. The fact that the hexagonally close-packed $Fe \cdot O \cdot OH$, goethite, provided a somewhat better fit to the EXAFS than the cubic close-packed lepidocrocite could be taken as support for the suggestion that loaded ferritin contains hexagonally close-packed arrays, perhaps related to haematite, goethite or a brucite-type mineral. This is broadly in agreement with the suggestion that the core contains ferrihydrite-like arrays, although it is stressed again that the uncertainties regarding the structure of this group of phases leave much to be desired.

3.4.5
Criterion (e): The Magnetic Properties

For loaded ferritin anomalous magnetic behaviours are observed which are in some way dependent on the overall loading of the individual molecules and also the nature of the mineralised core (for example, whether phosphate is present) [3]. The available magnetic data suggest that the coupling within ferritin cores is antiferromagnetic in nature [64], but that there are uncompensated spins on the surface of the cluster which contribute to a total ground state spin in a way analogous to ferrimagnetic interactions in extended lattices.

The magnetic properties of the cluster aggregates proposed as models for loaded ferritin indicate that in many cases the boundary effects of the particle play an important role. The elucidation of these properties is a complicated task, particularly for the larger clusters, and in some cases only very preliminary data have been reported. The most interesting magnetic properties arise in aggregates where there are some strong antiferromagnetic interactions between high spin iron(III) ions. In some cases, this leads to minimum ground state spins, for example, $S=0$ for Fe_{10}(can) [43]. In other cases, such as the Fe_8 [65], Fe_{10}(repw) [44] and Fe_{17}(crust)/Fe_{19}(crust) [56] the clusters carry high ground state spins as a consequence of unequal magnitudes in the antiferromagnetic interactions between pairs of iron(III) ions. These "spin frustration" phenomena lead to intermediate ground state spins and such clusters can be described as "ferrimagnetic" in a similar way to the ferritin case [66]. The Fe_8 and Fe_{10}(repw) systems appear to carry ground state spins in the region of $S=10$. The Fe_{17}(crust) and Fe_{19}(crust) crystallise within the same unit cell, complicating the interpretation of the magnetic data, but this system has been studied in depth, with the conclusion that at least one of the clusters stabilises a ground state spin of 33/2 or more, which is the highest value yet found for a molecular cluster. The mixed-valent Fe_{12}, molecule does not appear to be very strongly coupled, but Mössbauer data show certain parallels with the data recorded for animal ferritins [48]. Antiferromagnetic interactions leading to minimum ground state spins appear to dominate for Fe_{11} [51], Fe_{17}(sjl) [46] and Fe_{17}(repw) [45], with the Mössbauer data on Fe_{17}(sjl) indicating a superparamagnetic behaviour for this system [46].

Particles like the $Fe_{17/19}$(crust) clusters contain encapsulated portions of identifiable iron oxyhydroxide minerals, but it is clear that their magnetic behaviour is different from the "parent mineral". This is a consequence of the small size of the particle. We can define such clusters as zero-dimensional systems in which the original structure is reduced in all directions to such an extent that the properties of the original bulk system cannot be maintained and dramatic changes in physical properties are implied. For example, the particle may be too small to stabilise a magnetic domain, which is an alternative way of considering spin frustration effects and the stabilisation of intermediate spin states, these being partly a consequence of the strong boundary effects in small particles [67].

The superparamagnetic and recently reported quantum tunnelling effects in loaded ferritins are both consequences of the nanoscale nature of the iron core particles [68 and references therein]. As was pointed out earlier in Sect. 2.3, we are not at present in a position to know whether loaded ferritin cores contain one large particle or a number of smaller particles within the cavity. This complicates the interpretation of magnetic data, and it is to be hoped that investigations on smaller well-defined aggregates will shed light on the physical properties of ferritin cores. Clearly, the magnetic behaviour of the models for loaded ferritins varies from cluster to cluster. The full interpretation of the behaviour for clusters stabilising intermediate spins is a huge task and demands a range of experiments to be performed. This is an area where much work needs to be done, but the results obtained so far indicate that these scale models can aid in identifying the important criteria favouring the stabilisation of residual spins on cluster particles. Important points to address here are why some clusters show very little coupling of spins and why some are simply antiferromagnetically coupled with minimum gound state spins. The current view is that this is critically dependent on the nature of the bridging oxygen groups: for example, the bridging mode, μ_2, μ_3, or μ_4, and whether they are oxides, hydroxides or alkoxides. Studies on dinuclear species suggest that the latter point is highly influential on the magnitude of the exchange coupling [69]. Clearly this could be useful in helping to identify the proportions of oxide, hydroxide and water in the mineralised iron core of ferritin. Current work on the $Fe_{17/19}$(crust) system suggests that superparamagnetic and resonant tunnelling effects may come into play at very low temperatures, paralleling the higher temperature situation for loaded ferritin [70]. An investigation of these properties along with a quantification of the important exchange pathways in such clusters should help in elucidating the magnetic behaviour of ferritin.

3.4.6 *Criterion (f): Towards Functional Models*

The ultimate test of any model system is whether it can also function in a manner analogous to the real system. In the case of ferritin, this would mean that a model should be capable of assimilating and releasing iron in a reversible fashion ideally in an aqueous environment. The systems most likely to show such behaviour would be the nanoscale micelles and iron-delivery drugs as well as the cluster aggregates. Whilst it is clear that all of these can trap iron in the form of an oxyhydroxide-type material the questions of release and reversibility are harder to answer. Clearly the iron-delivery drugs can give-up their stores and it is possible that the sugars can trap new stores of iron subsequently. The mechanics of this have not been investigated. Most of the cluster molecules are highly unstable in aqueous environments. The mixed-valent clusters Fe_{12} and Fe_{17}(sjl) have been proposed as models for the initial formation of the mineral core. This assumes that core formation involves addition or incorporation in some way of iron(II) ions to the mineral. Since the details of the mechanism of core formation are not known, these molecules provide useful models to test this theory and they might turn out to be good functional models for iron assimilation. Experiments on aqueous solutions of the $Fe_{17/19}$(crust) system reveal that these clusters are metastable. Although they are not very soluble in water, once they are redissolved they evolve into the dinuclear compound $[Fe(heidi)(H_2O)]_2$ and, presumably, free iron(III) [57]. Since the yield of the dinuclear compound is quantitative and it has a ratio of Fe:heidi^{3-} of 1:1, whereas the ratio in the $Fe_{17/19}$(crust) is 2:1, the excess iron must be released into the medium. A space-filling picture shows that the iron in the centre of the crust molecules is relatively accessible, and it is tempting to identify the pathway through the surrounding organic "doughnut" with the entry and exit channels seen in ferritin (Fig. 15). Continuing the analogy we would then identify the

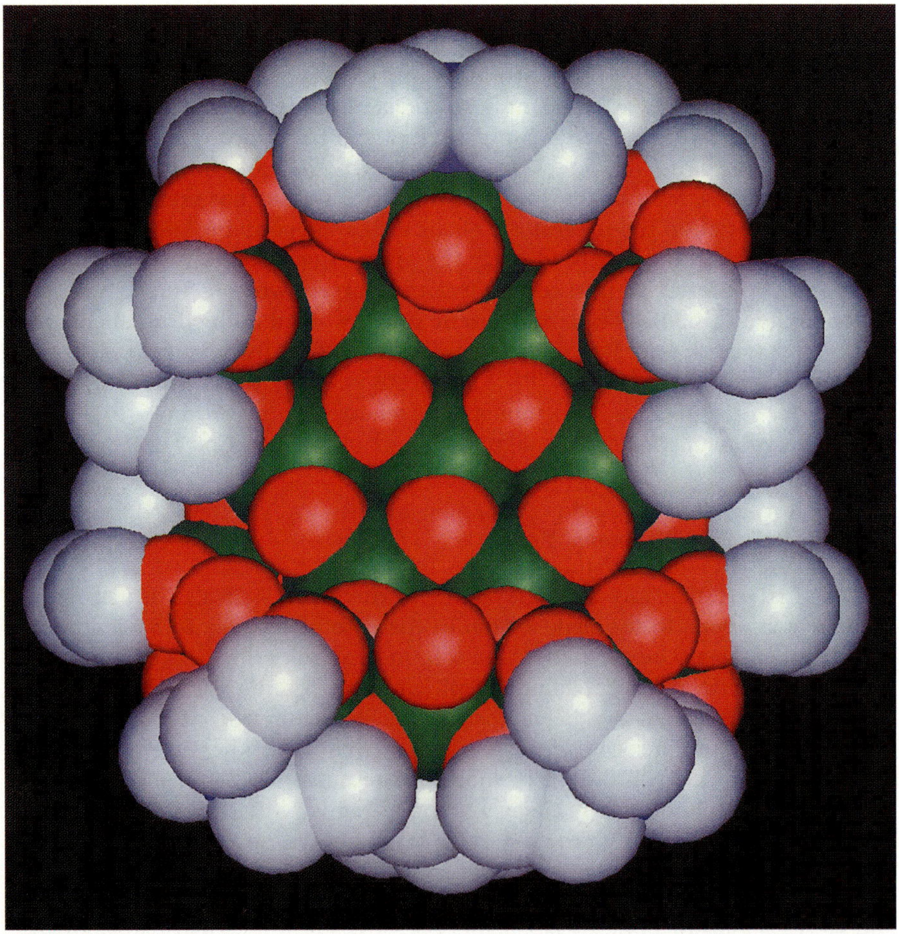

Fig. 15

heidi^{3-} ligands as corresponding to the polypeptide subunits of ferritin. Unfortunately, it is not possible to form an "apocrust" and the high stability of the dinuclear iron/heidi species means that crusts cannot be reformed under ambient conditions.

3.5 An Overall Assessment of the Iron Oxyhydroxide Clusters as Scale Models for Loaded Ferritin

Summarising the available data on these compounds, it can be seen that all of the clusters meet some of the suggested criteria quite well. At present, the clusters which come closest to meeting most of the criteria are the Fe_{17}(crust) and Fe_{19}(crust), which possess identifiable mineral cores; organic coats with the correct type of composition; have the right sorts of proportion of iron and core; reproduce the different iron environments measured for loaded ferritin using EXAFS closely; demonstrate anomalous magnetic behaviour; and can be made to release iron. It should be stressed that not all of the other proposed models have been tested on all of these points (particularly the last three). The other models with (pseudo) mineralised cores such as Fe_{10}(can) and Fe_{12} point to the directions which should be explored further. In particular, creating a three dimensional trapped mineral core is likely to provide insights into the situation in ferritin and should allow the relative proportion of oxide in the model to be increased, coming closer to the proposed scenario in loaded ferritin. This will probably be best achieved by engineering the ligand coat to favour three dimensional growth of the lattice. It should also be possible to incorporate other ligand "side-chains" in addition to the carboxylates, such as imidazoles. Those clusters containing more open iron core structures may well prove to be excellent models for partially loaded ferritin, particularly in terms of the relative composition of core and iron.

4 Conclusions and Outlook

In this review we have tried to show how the structures and properties of iron oxyhydroxide materials can shed light on the processes involved in producing biominerals. By concentrating on the mineralisation which occurs within ferritin we have been able to identify useful model systems which can further our understanding. A significant problem in identifying the mineral species contained within ferritin is that the size of the cavity in apoferritin places a restriction on methods such as X-ray diffraction for characterisation. This complicates the identification of the mineral type(s) within the core. The utility of extended mineral phases such as goethite as models for the core is limited since they cannot reproduce the variety of iron environments which will be present within the bounded nanoscale particles in loaded ferritin. Although iron oxyhydroxide particles which reproduce the dimensions of ferritin cores quite closely can be produced, these suffer from the same difficulties with characterisation as the ferritin cores themselves. Therefore, the most promising way forward is the use of scale models for loaded ferritins. At present these are relatively small cluster aggregates where iron oxyhydroxide units are encased in a coat of organic ligands. These models can reproduce many of the features of loaded ferritin in miniature. From the systems currently available it would seem that models containing recognisable portions of close-packed minerals provide the most suitable models. These systems also demonstrate most clearly the role of the organic coat in providing a template to direct crystal structure: an important feature to model in elucidating the processes involved in biomineralisation. In future, it should be possible to design organic templates which lead to larger, but still structurally characterisable, aggregates modelling the situation in loaded ferritin even more closely. These features include improving the relative composition of the core; improving the constitution of the organic coat to include more groups which mimic the amino acid sidechains present; enhancing the anomalous properties such as superparamagnetism and quantum tunnelling observed for ferritin; using the template ligand to stabilise different types of iron(III) oxyhydroxide structure in order to shed light on the identity of the mineral form in loaded ferritin; and lastly, creating models which can assimilate and release iron in a reversible manner.

Acknowledgements. The support of the BBSRC, EPSRC and Wellcome Trust is gratefully acknowledged by the author. Special thanks are due to Sigrid Wocadlo for help with the colour figures.

5 References

1. Harrison PM, Hoare RJ (1980) Metals in biochemistry, Chapman and Hall, London
2. Haber F, Weiss, J (1934) Proc Roy Soc Ser A 147:332
3. St. Pierre TG, Webb J, Mann, S (1989) Ferritin and hemosiderin: structural and magnetic studies of the iron core. In: Mann S, Webb J, Williams RJP (eds) Biomineralization. Chemical and biochemical perspectives. VCH, Weinheim, p 295
4. Frankel RB (1990) Iron biominerals: an overview. In: Frankel RB, Blakemore RP (eds) Iron Biominerals. Plenum, New York, p1
5. Flynn CM (1984) Chem Rev 84:31
6. Schneider W, Schwyn B (1987). In: Stumm W (ed) Aquatic surface chemistry. Wiley-Interscience, New York, p 167
7. Wells AF (1962) Structural inorganic chemistry, 3rd edn. Oxford University Press, London
8. Schneider W (1988) Chimia 42:9
9. Powell AK, Heath SL (1994) Comm Inorg Chem 15:255
10. Heywood BR, Mann S (1994) Chem Mater 6:311
11. Walsh D, Mann S (1995) Nature 377:320
12. Walsh D, Mann S (1996) Chem Mater 8:1944
13 Ford GC, Harrison PM, Rice DW, Smith JMA, Treffry A, White JL, Yariv J (1984) Phil Trans R Soc Lond B 304:551
14. Harrison PM, Arosio P (1996) Biochim Biophys Acta 1275:161
15. Lobreaux S, Yewdall SJ, Briat JF, Harrison PM (1992) Biochem J 288:931
16. Meldrum FC, Heywood BR, Mann S (1992) Science 257:522
17. Meldrum FC, Wade VJ, Nimmo DL, Heywood BR, Mann S (1991) Nature 349:684
18. Wade VJ, Treffry A, Laulhere J-P, Bauminger ER, Cleton MI, Mann S, Briat J-F, Harrison PM (1993) Biochim Biophys Acta 1161:91
19. Mann S, Williams JM, Treffry A, Harrison PM (1987) J Mol Biol 198:405

20. Ringeling PL, Davy SL, Monkara FA, Hunt C, Dickson DPE, McEwan AG, Moore GR (1994) Eur J Biochem 223:847
21. Le Brun NE, Thomson AJ, Moore GR (1997) Structure and Bonding 88:103
22. Moore GR, Mann S, Bannister JV (1986) J Inorg Biochem 28:329
23. Mann S, Bannister JV, Williams RJP (1986) J Mol Biol 188:225
24. Harrison PM, Clegg GA, May K (1980) Prog Biophys Mol Biol 36:131
25. Stumm W, Morgan JJ (1981). Aquatic chemistry. Wiley-Interscience, New York
26. Eggleton RA, Fitzpatrick RW (1988) Clays Clay Miner 36:111
27. Robie RA, Bethke PM, Beardsley KM (1980) X-ray crystallographic data, molar volumes and densities of minerals and related substances. In: Weast RC (ed) CRC Handbook of Chemistry and Physics, 61st edn. CRC Press, Florida, p B-208
28. Towe KM, Bradley WF (1967) J Colloid Inter Sci 24:384
29. Chukrov FV, Zvyagin BB, Gorschkov AI, Ermilova LP, Balashova VV (1973) Int Geol Rev 16:1131
30. Russell JD (1979) Clay Miner 14:109
31. Cornell RM, Giovanoli R, Schneider W (1989) J Chem Technol Biotechnol 46:115
32. Manceau A, Combes J-M, Calas G (1990) Clays Clay Miner 38:331
33. Eggleton RA, Fitzpatrick RW (1990) Clays Clay Miner 38:335
34. Brindley GW (1984) Order-disorder in clay mineral strcutures. In: Brindley GW, Brown G (eds) Crystal structures of clay minerals and their X-ray identification. Mineralogical Society, London p 222
35. Spiro TG, Allerton SE, Renner J, Terzis A, Bils R, Saltman P (1966) J Am Chem Soc 88: 2721
36. Spiro TG, Pape L, Saltman P (1967) J Am Chem Soc 89:5555
37. Mann S, Hannington JP, Williams RJP (1986) Nature 324:565
38. Mann S, Hannington JP (1988) J Colloid Interface Sci 122:326
39. Theil EC, Sayers DE, Brown MA (1979) J Biol Chem 254:8132
40. Yang C, Bryan AM, Theil EC, Sayers DE, Bowen LH (1986) J Inorg Biochem 28:393
41. Hagen KS (1992) Angew Chem Int Ed Engl 31:1010
42. Lippard SJ (1988) Angew Chem Int Ed Engl 27:344
43. Caneschi A, Cornia A, Fabretti AC, Gatteschi D (1995) Angew Chem In Ed Engl 34:2716
44. Benilli C, Parsons S, Solan GA, Winpenny REP (1996) Angew Chem Int Ed Engl 35:1825
45. Parsons S, Solan GA, Winpenny REP (1995) J Chem Soc Chem Commun 1987
46. Micklitz W, McKee V, Rardin RL, Pence LE, Papaefthymiou GC, Bott SG, Lippard SJ (1994) J Am Chem Soc 116:8061
47. Taft, KL, Papaefthymiou GC, Lippard SJ (1993) Science 259:1302
48. Taft, KL, Papaefthymiou GC, Lippard SJ (1994) Inorg Chem 33:1510
49. Micklitz W. Lippard SJ (1989) J Am Chem Soc 111:6856
50. Gorun SM, Lippard SJ (1986) Nature 319:666
51. Gorun SM, Papaefthymiou GC, Frankel RB, Lippard SJ (1987) J Am Chem Soc 109:3337
52. Taft KL, Lippard SJ (1990) J Am Chem Soc 112:9629
53. Wieghardt K, Pohl K, Jibril I, Huttner G (1984) Angew Chem Int Ed Engl 23:77
54. Hegetschweiler K, Schmalle H, Streit HM, Schneider W (1990) Inorg Chem 29:3625
55. Heath SL, Powell AK (1992) Angew Chem Int Ed Engl 31:191
56. Powell AK, Heath SL, Gatteschi D, Pardi L, Sessoli R, Spina G, Del Giallo F, Pieralli F (1995) J Am Chem Soc 117:2491
57. Heath SL (1992) PhD Thesis, UEA, Norwich, UK
58. Deer WA, Howie RA, Zussman J (1965) Rock forming minerals, vol. V, non-silicates, 4th edn. Longmans, London
59. Heath SL, Charnock JM, Garner CD, Powell AK (1996) Chem Eur J 2:634
60. Islam QT, Sayers DE, Gorun SM, Theil EC (1989) J Inorg Biochem 36:51
61. Mackle P, Garner CD, Ward RJ, Peters TJ (1991) Biochim Biophys Acta 1115:145
62. Heald SM, Stern EA, Bunker B, Holt EM, Holt SL (1979) J Am Chem Soc 101:76
63. Rohrer JS, Quazi TI, Watt GD, Sayers DE, Theil EC (1990) Biochem 29:259
64. Granick S (1945) Chem Rev 38:379

65. Delfs CD, Gatteschi D, Pardi L, Sessoli R, Wieghardt K, Hanke D (1993) Inorg Chem 32:3099
66. Gatteschi D, Caneschi A, Pardi L, Sessoli R (1994) Science 265:1054
67. de Jongh LJ (1994) Physics and chemsitry of metal cluster compounds, Kluwer Academic Publishers, Dordrecht
68. Tejada J, Ziolo RF, Zhang XX (1996) Chem Mater 8:1784
69. Kurtz DM (1990) Chem Rev 90:585
70. Ballou R, Lionti F, Powell AK unpublished work

Heme: The Most Versatile Redox Centre in Biology?

Stephen K. Chapman, Simon Daff and Andrew W. Munro

Department of Chemistry, University of Edinburgh, West Mains Road, Edinburgh EH9 3JJ, U.K. *Email: S.K.Chapman@ed.ac.uk*

Iron-porphyrin complexes, known as hemes, form the prosthetic groups of a number of proteins. These hemoproteins exhibit an impressive range of biological functions. These include: Simple electron transfer reactions, oxygen transport and storage, oxygen reduction to the level of hydrogen peroxide or water, oxygenations of organic substrates, and the reduction of peroxides. This diversity of function is often extended further by combining heme groups with other cofactors, e.g. flavins and/or metal ions. Such combinations frequently allow heme cofactors to couple electron transfers with other processes, such as the translocation of protons or the reduction/oxidation of other molecules. This versatility in function is made possible by a combination of differences in both the polypeptide and heme constituents of the various hemoproteins. The aim of this article is to illustrate how nature has used different protein frameworks to exploit the redox properties of heme. This is done by focusing on a carefully chosen selection of hemoproteins which exemplify the numerous redox functions performed by heme in biology.

Keywords: Heme; cytochromes; electron transfer; P-450; oxidases; oxygen; monooxygenases; peroxidases; flavocytochromes; redox centres

1 Introduction

The title of this article is deliberately provocative. It might be challenged in a number of ways depending on your standpoint. An organic chemist might argue that a species such as the isoalloxazine ring of a flavin can fulfil many of the functions performed by heme in nature. Similarly, it might be suggested that metal ions such as copper, when appropriately bound in a protein, can accomplish most of the redox processes carried out by heme. So, the question arises, is heme really such a versatile cofactor in biology? We will address this question by considering the various biological activities facilitated by heme groups and illustrating these with one or more carefully chosen examples of hemoproteins which perform these particular functions.

Heme-containing proteins display a diversity of biological functions. These include: Simple electron transfer reactions, such as those catalysed by b- and c-type cytochromes [1], oxygen transport and storage via hemoglobin and myoglobin [2], oxygen reduction to the level of water by cytochrome oxidase [3], oxygenations of organic substrates as facilitated by the cytochromes P-450 [4] and the reduction of peroxides by the catalases and peroxidases [5]. This range of activities can be extended by combining heme groups with other cofactors, such as flavins and/or metal ions, such as molybdenum or copper. Such combinations allow heme cofactors to participate in a range of enzymatic processes, including dehydrogenations [6] and the reduction of a number of small molecules [7].

This flexibility in function arises from a combination of differences in both the polypeptide and heme constituents of the various hemoproteins. As far as the heme group is concerned, there are four common representatives found in biology and these are illustrated in Fig. 1. Of these, heme-b (also termed protoheme IX) is perhaps the most familiar, since this forms the prosthetic group of hemoglobin, myoglobin, catalase, most peroxidases, the b-type cytochromes and the cytochromes P-450. Heme-c differs from heme-b by being covalently linked to the protein via thioether linkages which form between the thiol groups of cysteine residues on the protein and the 1′-carbons of the vinyl groups at positions 2 and 4 of the porphyrin ring. Heme-c is of course the prosthetic group for the c-type cytochromes [8–10], which are probably the most extensively studied family of electron transfer proteins. Heme-*a* provides cofactors for the cytochrome aa_3 of cytochrome oxidase in which one *a*-type heme acts as a simple electron transfer agent and the other provides a dinuclear site, along with Cu_B, at which oxygen reduction can take place [3]. In heme-*a* there is a hydroxyethylfarnesyl sidechain at position 2 of the porphyrin ring and a formyl group at position 8. Heme-d_1 is found as a prosthetic group in the heme-containing nitrite reductases designated as cytochromes cd_1 [11]. In addition to differences in sidechains, heme-d_1 differs from hemes *a*, b and c by having saturated bonds in the ring. The presence of four carboxylic acid groups in heme-d_1 make it the most water soluble of the heme groups and it is easily extracted from the cd_1 enzymes by acid treatment [12].

In addition to the heme group itself, the presence and the nature of fifth and six ligands to the iron will be crucial in determining the reactivity of any hemoprotein. Having a coordination site available to bind exogenous ligands is obviously of paramount importance if, for example, the hemoprotein is an oxidase or oxygenase. In contrast, the hemoproteins involved in simple electron

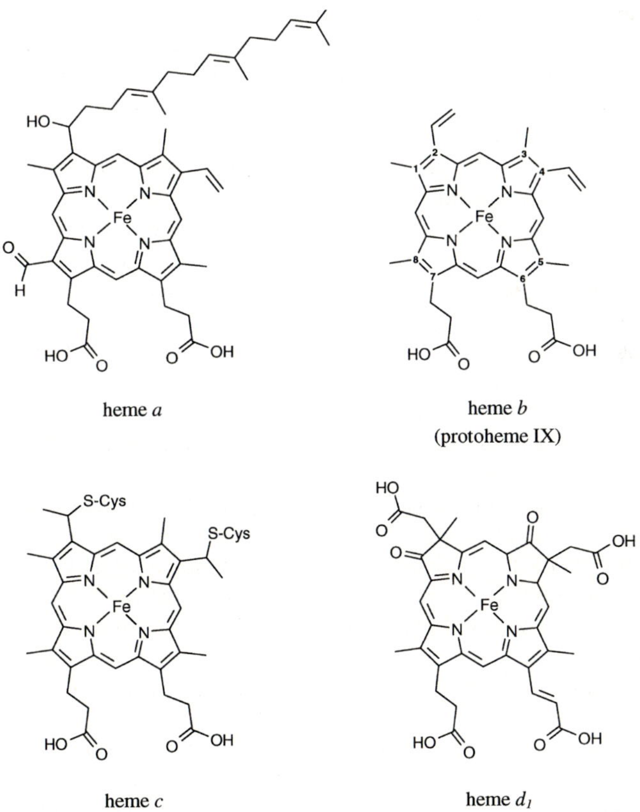

Fig. 1. Hemes commonly found in biological systems

transfers need to avoid substantial reorganisation at the iron site and are, therefore, almost exclusively hexa-coordinated with the iron in the low-spin state. The different ligation schemes and their effects on the properties of hemoproteins have been clearly and comprehensively reviewed previously by Moore and Pettigrew [10].

The huge interest in heme-containing proteins has generated an extensive literature on the subject with many excellent reviews covering the various protein types in detail [1–11]. The purpose of this article is not to exhaustively revisit what is already covered in these reviews. Instead the aim is to illustrate the variation in the biological function of hemoproteins by using just a few, hopefully well chosen, examples.

2 Electron Transfer

Biological electron transfer is used in nature primarily for energy transduction, for example in photosynthesis or in respiration. These processes are facilitated by a series of individual proteins, often membrane-bound, which provide what might be described as the electrical "wiring" of the cell. Many of the proteins which form this wiring are heme-containing and are known as cytochromes.

Biological redox reactions vary in their complexity. They might, for example, involve electron transfer coupled to atom and/or ion transfer, and such reactions are illustrated in the case of heme-containing oxidases and monooxygenases in Sects. 3.2 and 3.3. They may, on the other hand, be relatively simple as exemplified by a one-electron transfer from one cytochrome to another. It is these types of redox reactions that form the focus of this section.

The basis of our understanding of the factors which control such electron transfers is principally based on the theoretical analysis provided by Marcus [13–16]. The elements of this theory and the way in which it applies to biological systems have been the subject of intense interest and there are now numerous reviews focusing on protein mediated electron transfer [17–21]. The recent book on protein electron transfer edited by Bendall is particularly recommended in this context [22].

In physiological systems, it is generally advantageous for an electron transfer reaction to maximise the driving force (ΔG) and minimise the reorganisation energy (λ). Such conditions will provide the most efficient electron transfer and thus the fastest rates (although in extreme cases where the reaction is "overdriven", i.e. the 'Marcus Inverted Region', the opposite is true [17]). Large reorganisation energies are associated with rearrangements at an atomic level and tend to be lower in rigid hydrophobic systems. Therefore, in cytochromes associated with simple electron transfers the heme is almost always hexa-coordinated, low spin, and bound within a hydrophobic crevice. The extended π-system of the porphyrin ring delocalises the electron over a wide area and reduces the necessity for re-ordering both in the local structure and in the surrounding solvent molecules. The driving force is determined by the redox potential of the donor and acceptor, and, as discussed later, this is modulated by the surrounding protein. Moser et al. [17] showed that, within biological systems, electron transfer rates generally decrease exponentially with the distance between the cofactors at an average rate of 1.4 $Å^{-1}$. However, the possibility of distinct pathways acting as faster transfer routes should also be considered [23].

It would appear that nature controls electron transfer reactions simply by modulating distance and reduction potential. During respiration and photosynthesis, each electron transfer reaction represents a waste of metabolic energy, as the high energy electron falls further into a potential well. Therefore, energetic efficiency must be balanced against speed in the selection of these parameters.

2.1 Modulating Reduction Potentials

The most straightforward role of heme in biology is exemplified by the class I cytochromes c which act simply as single electron transfer mediators. The most common and most studied of these is mitochondrial cytochrome c, which acts as the electron donor to cytochrome c oxidase (aa_3) at the end of the respiratory chain (see Sect. 3.2). The reduction potential of mitochondrial cytochrome c is relatively high, corresponding to its penultimate position in the respiratory electron transport chain. However, the class I cytochromes c are quite diverse and have adapted to a variety of biological environments. Accordingly, the versatility of the heme redox cofactor has been exploited by manipulating the reduction potential to suit these alternative environments (see Table 1).

Table 1. Midpoint reduction potentials for a selection of simple cytochromes

Protoheme IX [24]	–	– 115
Class I cytochromes c		
Mitochondrial cytochrome c	His-Met	+ 260
– mutant M80H [25]	His-His	+ 41
M. braunii cytochrome c_6 [26]	His-Met	+ 358
D. vulgaris cytochrome c-553 [27]	His-Met	~ 10
Algal cytochrome c-550 [28]	His-His	– 260
D. vulgaris cytochrome c_3	His-His	– 280
	His-His	– 320
	His-His	– 380
	His-His	– 80
– mutant H70 M [30]	His-Met	– 80
R. viridis tetraheme cytochrome [31]	His-Met	+ 380
	His-His	+ 20
	His-Met	+ 310
	His-Met	– 60
Bacterioferritin [32]	Met-Met	– 225
+ ferric core		– 475

2.1.1
Axial Ligation

As demonstrated in Table 1, the reduction potential of a heme group can be modulated over a range of more than 0.7 V by the surrounding protein matrix. Clearly, the nature of the axial ligands is an important consideration and in electron transfer proteins these are usually either histidine or methionine or a combination of both (exceptions include cytochrome *f* in which the iron is ligated by a histidine and the amine group of the protein *N*-terminus). In general, it is found that among c-type cytochromes His-His ligation correlates with lower reduction potentials than His-Met, although this is not always the case.

Mutagenesis studies have allowed the importance of ligand choice to be quantified in a more or less fixed environment. The two mutants represented in Table 1 switch the axial ligands of two proteins, one from His-Met to His-His [25] and the other from His-His to His-Met [30]. The result, in both cases, is a reduction potential change of around 200 mV, which indicates crudely the degree of influence exerted by the choice of ligand. An interesting natural parallel can be drawn with cytochrome c_{550} from algae. This has a very low reduction potential, for a class I c-type cytochrome, which appears to have been caused by a His-Met to His-His ligand change during evolution [28, 29]. Despite its different ligation, it can still be considered to be a class I cytochrome through its sequence similarity to cytochrome c_6. *D. vulgaris* cytochrome c_{553}, on the other hand, maintains a low reduction potential without resorting to a ligand change [27]. It is clear, therefore, that although the choice of ligand is important, other structural factors must come into play.

2.1.2
Heme-Heme Interactions

One particular point of interest in multi-heme enzymes, such as the tetraheme cytochromes, is the effect of the oxidation state of neighbouring hemes. The sequential reduction/oxidation of hemes in such enzymes leads to standard reduction potentials which are termed "macroscopic". However, attempts have also been made to separate the microscopic events, in order to quantify the dependency of the reduction potential of each heme on the variable electrostatic field caused by the neighbouring ones. Cytochrome c_3, a tetraheme class III cytochrome c found in sulfate-reducing bacteria, has been used as a model for two such studies [33, 34]. The heme-heme interactions were found to account for differences of up to 60 mV between the macroscopic and microscopic potentials. This is clearly a significant factor in systems where all the potentials can fall within a spread of 100 mV.

Most of the controlling factors in determining heme reduction potentials are represented in the tetraheme cytochrome from the *R. viridis* reaction centre. This has three His-Met ligated hemes and one His-His. Surprisingly, the His-His heme has a higher E_m than one of the His-Met ligated variants. The wide variation in the heme potentials made this protein an ideal target for theoretical modelling. Gunner and Honig [31] attempted this by summing a series of predefined factors to account for axial ligation, electrostatic interactions, the charge of heme propionates and the oxidation state of neighbouring hemes. The results show a broad correllation, correctly predicting the large difference between the two low-potential hemes and the two high-potential ones. However, within these pairs differentiation was less successful. Clearly, to predict reduction potentials accurately, all aspects of the protein matrix must be accounted for (see[1]), as must additional complications, such as the presence of multiple cofactors. For example, the presence of the ferric core shifts the reduction potential of the bacterioferritin heme from –225 mV to –475 mV [14], an effect which may help to regulate the kinetics of the iron storage process (see Sect. 2.2.3.3).

2.2
Directing Electron Transfer

Biological electron transfer occurs down a potential well, the depth of which is determined by the reduction potentials of the cofactors. However, to maximise the rate of transfer the distance between cofactors must be minimised. This is achieved in membrane-spanning systems, such as cytochrome c oxidase (see Sect. 3.2) and the photosynthetic reaction centre, by fixing several redox sites close together within a rigid protein framework. Alternatively, soluble redox proteins use features such as electrostatics and surface complementarity to encourage the formation of specific bimolecular complexes. The study of these protein:protein interactions has been aided recently by the availability of X-ray structures derived from the co-crystals of physiological complexes [35].

2.2.1
Cytochrome c: Cytochrome c Peroxidase

The complex between the physiological redox partners cytochrome c and cytochrome c peroxidase, formed by co-crystallizing the two proteins, illustrates the important factors in interprotein electron transfer (see Fig. 2) [36]. Cytochrome c has been shown to utilise a number of positively charged surface residues when interacting with its redox partners. These surround the exposed edge of the heme to ensure close contact between this and the redox partner on complexation. In the complex, interactions between cytochrome c Lys5, Lys87, Lys73 and cytochrome c peroxidase Asp34, Glu35, Glu290 appear to be important, an observation backed up by kinetic analysis of a number of point mutants [37]. Detailed examination of the complex, however, indicates that most of the interactions in the complexes are hydrophobic in nature [36]. An electron transfer-tunnelling pathway (Fig. 3) has also been proposed, which travels from the surface residue Ala193 (which contacts the exposed heme methyl group of cytochrome c) along the peptide chain to the redox-active residue Trp191 (see Sect. 3.4.2). Mutation of Ala193 to Phe disrupts the electron transfer event, presumably by increasing the distance between the redox centres. Further evidence in support of this location as the cytochrome c electron transfer binding

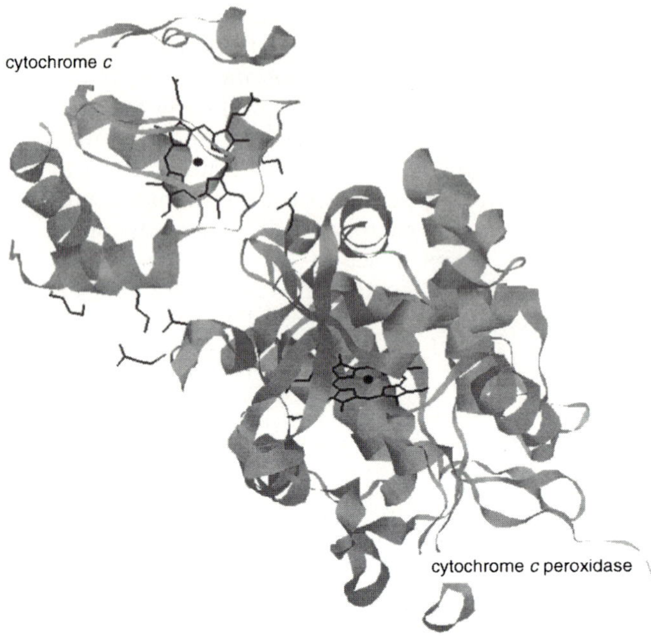

Fig. 2. The complex formed between yeast cytochrome c and cytochrome c peroxidase on crystallisation [36]. The two heme groups are highlighted along with three salt bridges believed to contribute to stabilisation of the complex (cc:ccp – Lys5:Glu35, Lys73:Glu290, Lys87: Asp34)

site was obtained by covalently attaching cytochrome c to the peroxidase via engineered cysteine residues [38, 39]. Attaching the protein at the proposed binding site (according to the crystal structure) was found to inactivate the enzyme with respect to reduction by ferrocytochrome c. There is some evidence for the presence of a second, low-affinity binding site, which may utilise a more efficient electron transfer route directly to the heme cofactor [40–42]. However, recent experiments have cast some doubt upon this hypothesis [38, 39].

2.2.2
Cytochrome c_6 vs Plastocyanin

Cytochrome c_6 (or c-553; a high potential type I cytochrome) acts as the ultimate electron donor to photo-oxidised chlorophyll in photosystem I of some algae and cyanobacteria. Therefore, its role can be compared directly with that of the blue copper protein, plastocyanin, in plants. The comparison is particularly relevant in the unusual species of algae and cyanobacteria which contain both enzymes [43, 44], where valuable insight can be obtained into the important factors governing the selection of a particular electron transferase by an organism and, generally, into the evolution of electron transfer complexes. In

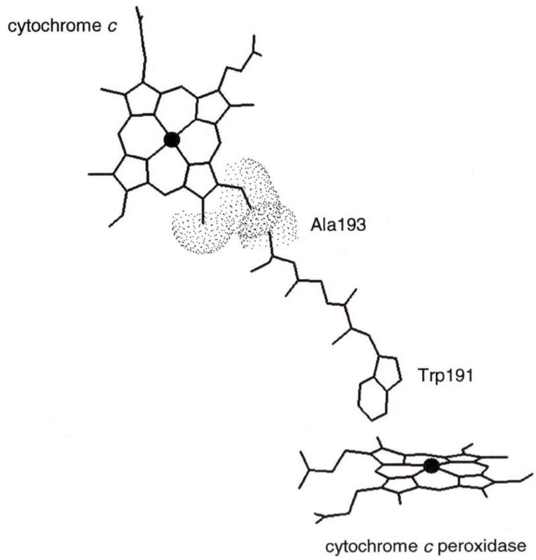

Fig. 3. The putative electron transfer pathway from the cytochrome c heme to the cytochrome c peroxidase heme passes from the interface residue Ala193 to the Trp191 redox centre via the peptide backbone [36]

such organisms, the proportion of each protein expressed probably hinges on the availability of copper. This supports the view that cytochrome c_6 has been superseded by plastocyanin in higher organisms. Similarities in the kinetic behaviour and physical characteristics of enzymes from the same source have been noted [45, 46]. These can now be interpreted in respect of X-ray crystal [47–49] and NMR solution [50] structures which are available for both enzymes, albeit from different sources. In the locality of the exposed heme group, the cytochrome possesses hydrophobic and negative regions which appear to correlate with the putative electron transfer site on plastocyanin. Presumably, the surfaces have evolved in parallel to complement those of their shared redox partners. However, it has also been noted that plastocyanin possesses two distinct sites specific for its two redox partners (photosystem I and cytochrome *f*), whereas cytochrome c_6 appears to have only one [49, 50]. The asymmetrically positioned heme enclosed in an α-helix sandwich is a motif common to all class I cytochromes c and limits the electron transfer site to a single face of the protein, at which the heme edge is exposed. Therefore, its binding mode is unlikely to differ greatly from that of mitochondrial cytochrome c illustrated in Fig. 2. This contrasts with the complexes formed by blue copper proteins. A particularly good example is the ternary complex formed between amicyanin and its two redox partners (cytochrome c-551i and methylamine dehydrogenase) crystallised by Chen et al [51]. It is intriguing that two such different electron transferases can evolve characteristics similar enough to allow direct substitution for one another so readily *in vivo*.

2.2.3
Coupled Systems

In addition to their ubiquitous presence as simple electron transfer mediators, hemes can also be found in enzyme systems which link electron transfer to other functions. The coupling mechanism can be both direct, as in the case of the cytochromes P-450 and peroxidases, or indirect as in the case of cytochrome c oxidase. These complex coupled systems are covered in detail in Sects. 3.2, 3.3 and 3.4.

2.2.3.1
Flavocytochromes

Simple cytochromes used universally as electron mediators are also found fused to functional enzymes to allow direct coupling of two-electron to one-electron redox processes (i.e. they can function as biological transformers). A particularly well characterised example is flavocytochrome b_2 from yeast [6, 52]. This is a homotetrameric enzyme composed of an FMN-containing dehydrogenase domain and a monoheme b-type cytochrome domain (Fig. 4). It couples the dehydrogenation of L-lactate to the reduction of mitochondrial cytochrome c which, in turn, leads to the generation of a single equivalent of

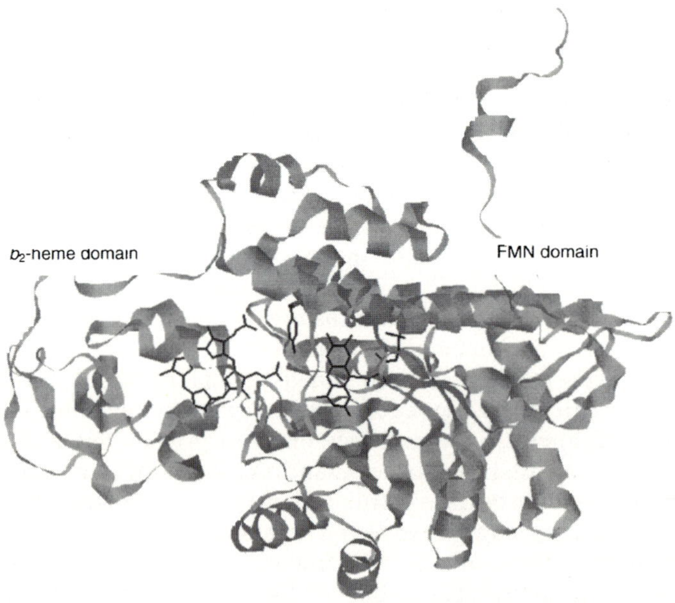

Fig. 4. Flavocytochrome b_2 from *S. cerevisiae* comprises a 400 residue FMN binding domain which is connected to a 100 residue heme-binding domain via a short polypeptide "hinge". In the high resolution crystal structure, Tyr143 from the FMN domain (highlighted) can be seen to hydrogen bond to one of the b_2-heme propionates in one crystallographically distinguishable subunit, while it hydrogen bonds to the substrate in another [58]

ATP [53]. In the absence of the cytochrome domain, cytochrome c reductase activity is virtually lost [54], whereas the lactate dehydrogenase activity is unaffected. The role of the heme domain is therefore unambiguous. A number of mutagenesis studies have been performed to locate the important interdomain contacts, but only two significant regions have been found. Firstly, the length of polypeptide, or "hinge", connecting the two domains appears to lock the domains together in the ideal conformation for electron transfer [55–57]. Secondly, a tyrosine residue (Tyr143) was found to H-bond to one of the heme propionates in one crystallographically distinguishable subunit and to the substrate in the other [58]. This was interpreted as evidence for its use as an electronic switch during turnover, converting the enzyme from a dehydrogenase to an electron transferase. Mutation of this residue to Phe disrupted flavin-to-heme electron transfer [59]. There are two flavin-to-heme electron transfers in the catalytic cycle; firstly flavin hydroquinone to heme and secondly flavin semiquinone to heme. The second of these electron transfers has been found to be the main rate-determining step of the catalytic cycle [60]. The transfer of electrons from the b_2-heme domain to cytochrome c appears to pass mainly through one specific site, although cytochrome c reduction probably occurs at a decreased rate from alternative sites when the primary site is occupied [61]. The credibility of this situation is supported by an appraisal of the electron transferase ability of the enzyme [62].

Several other flavocytochromes exist and perform a variety of tasks, where the heme groups act as either electron acceptors from flavin or electron donors to it. An example of the latter variety is flavocytochrome c_3 from *Shewenella putrefaciens*, which acts as a fumarate reductase [63, 64]. This enzyme appears to have evolved from a fusion of a reductase domain to a tetraheme cytochrome, which is related by sequence to the class III c-type cytochromes (i.e. cytochromes c_3) [65]. The heme groups have characteristically low reduction potentials [63] making them appropriate electron donors. However, the roles of cytochromes c_3 in general are still ambiguous [66].

2.2.3.2
Cytochrome cd_1

In cytochrome cd_1, electron transfer is coupled to the one-electron reduction of nitrite to nitric oxide and, alternatively, to the reduction of dioxygen to water. It is the only heme d_1 utilising enzyme in which the cofactor is non-covalently bound. The high resolution X-ray crystal structure for this enzyme has clarified a number of contentious points and has allowed the authors to propose a detailed mechanism for the reaction [67]. The enzyme is a homodimer with each subunit comprising a small c-heme domain (His-His ligated) and a larger d_1-heme domain (His-Tyr ligated). The c-type heme has been shown to function as an internal electron transfer mediator between the physiological substrate, which is probably cytochrome c_{551}, and the d_1-heme [1, 68]. Its removal by controlled proteolysis eliminates this aspect of the enzyme's function, while the remaining d_1-domain is still able to function as an oxidase [69]. However, the rate of electron transfer from heme c to heme d_1 has been shown to rate-limit turnover to less than 1 s^{-1}, despite an edge-to-edge inter-heme distance of only 11 Å [70,71]. There is some evidence (from EPR and MCD studies) that the d_1-heme undergoes a spin-state change on reduction, resulting from an alteration in the ligand field of the iron. Fulop et al. [67] propose that the axial ligand Tyr25 dissociates from the metal allowing the substrate (nitrite) to co-ordinate in its place. Coupling between this large rearrangement and d_1-heme reduction could be responsible for the low rate of electron transfer observed. Intriguingly, the Tyr25 ligand is offered, not from the d_1-heme domain, like the other axial ligand His200, but from the c-heme domain. In fact, it is connected by a short strand of polypeptide to His17, which is an axial ligand to the c-heme. Therefore, at the end of the catalytic cycle, when NO is bound to Fe(III) at the d_1-heme, Tyr25 is in position to displace simultaneously the product and draw the hemes together to initiate electron transfer. This is considered to be a key aspect of the mechanism, since it prevents the formation of the stable Fe(II)-NO adduct which could act as a 'dead end' complex [70].

2.2.3.3 Ferritins

The ferritins are a particularly unusual case, since electron transfer appears to be coupled to the storage and release of non-heme iron via a b-type heme. The crystal structure of bacterioferritin [72] shows it to be a 24-subunit sphere encompassing a cavity some 80 Å in diameter. At each inter-subunit contact, a heme b-557 (12 in all) is ligated via two methionines (one from each subunit). Iron (II) is complexed by the enzyme at non-heme iron sites (one per subunit) before being deposited in the central cavity as iron (III) [73a]. It is interesting to note that iron uptake still occurs in the absence of heme [73b]. On the other hand, evidence suggests that bound heme does aid the release of iron (II) from the iron (III) core [73c]. The identity of the physiological redox partner for ferritin is central to the ongoing discussion regarding its exact role. Recently, a 64 residue [2Fe-2S] protein has been expressed from a gene located upstream of the bacterioferritin gene in *E. coli* and purified. It was found to bind specifically to bacterioferritin, although the complex remains uncharacterised [74].

3 Dealing with Oxygen

This section will focus primarily on the heme-containing enzymes involved in the activation and reduction of dioxygen. Hemoproteins involved in what are formally non-redox roles, e.g. transport and storage of dioxygen (i.e. hemoglobin and myoglobin), etc., are deliberately excluded.

3.1 Dioxygen and Its Reduction Products

Dioxygen, O_2, is thermodynamically a powerful oxidising agent, yet simple bimolecular reactions with organic substrates are not common. Indeed, reactions of O_2 with organic compounds usually require some kind of activation by radical promotion, metal ion catalysis, or photochemical excitation. The kinetic inertness of dioxygen arises, to a large extent, from the electronic nature of the molecule, which exists in a triplet ground state. Thus, although both the two-electron reduction of O_2 to H_2O_2 and the four-electron reduction of O_2 to $2H_2O$ are thermodynamically very favourable (the reduction potentials being +0.281 V and +0.815 V, respectively, versus NHE at pH 7.0 and 25 °C) they are both spin-restricted, since O_2 has a triplet ground state whereas H_2O_2 and H_2O have singlet ground states. The one-electron reduction to superoxide, on the other hand, is thermodynamically unfavourable with a reduction potential of -0.33 V.

Although dioxgen is inert towards organic species, it does bind to and react very readily with transition metals [75]. It is not surprising, therefore, to find that when oxygen reduction occurs in biology, it is usually accomplished by metal ions (exceptions include the many two-electron oxidases which have flavin cofactors [76]).

In biological systems the dioxygen molecule is used in a number of ways. It can act as a terminal electron acceptor, as a metabolite and even as a toxin. When O_2 acts as a terminal electron acceptor, it is reduced either to H_2O_2 or H_2O by enzymes known as oxidases. Oxidases have three substrates; molecular oxygen, protons and a reducing substrate. A good example to illustrate the role of heme in oxygen reduction is the enzyme cytochrome c oxidase and this is the focus of Sect. 3.2.

In addition to its use as an electron acceptor, O_2 is also used directly in the synthesis and degradation of many of the chemical constituents of the cell. The enzymes which catalyse such reactions are termed oxygenases. These enzymes either insert one O-atom of dioxygen into a substrate (reducing the other to water) in which case they are monooxygenases, or insert both O-atoms into a substrate in which case they are dioxygenases. The classic group of heme-containing oxygenases are the cytochromes P-450 and these are discussed in Sect. 3.3.

3.2 Oxygen as a Terminal Electron Acceptor: Cytochrome c Oxidase

One of the most complex and important of all heme-containing proteins is cytochrome c oxidase [77]. The last few years have seen major breakthroughs in our understanding of this enzyme. In 1995 the three-dimensional structures of cytochrome c oxidases from *Paracoccus denitrificans* [78] and bovine heart [79] were reported. These beautiful structures have, at last, allowed the function of the enzyme to be understood in the context of a firm structural framework.

Mammalian cytochrome c oxidase is a complex membrane-bound protein composed of thirteen subunits (there are only four in the simpler bacterial enzyme [78]) with an overall M_r of around 200 kDa. The enzyme spans the mitochondrial innermembrane where its function is to transfer electrons from cytochrome c, located in the inter-membrane space, to molecular oxygen in the matrix [80–82]. These electron transfers are coupled to the translocation of four protons across the membrane from the matrix to the inter-membrane space. [3, 83–85]. This movement of protons generates a proton motive force across the membrane which is used to drive the synthesis of ATP from ADP via the F_1F_0 ATP synthase [86]. When the four protons required for water formation are also taken into account, it can be seen that each electron transfer results in the uptake of two protons from the matrix; one translocated and the other used in H_2O production. Thus, the overall reaction catalysed by cytochrome c oxidase is shown below:

$$4\ \text{Cyt}\ c^{2+} + O_2 + 8H^+_{[in]} \rightarrow 4\ \text{Cyt}\ c^{3+} + 2H_2O + 4H^+_{[out]}\ .$$

(Where: $H^+_{[in]}$ = protons in the matrix; $H^+_{[out]}$ = protons pumped out of the matrix; Cyt c^{2+} = reduced cytochrome c; Cyt c^{3+} = oxidised cytochrome c).

Therefore cytochrome c oxidase can be thought of as a proton pump driven by electron transfer [3, 78–85, 87]. Suggestions have been made concerning the actual pathway in the protein through which the protons are pumped [78, 88]. Possible mechanisms for proton translocation include the suggestion that the imidazole groups of histidine residues shuttle between protonation states in a so called "histidine cycle mechanism" [10, 89]. It has also been suggested that one of the metal centres in the enzyme, Cu_B, along with a particular ligand residue, His325, may also play a role in controlled proton translocation.

3.2.1
The Redox Centres in Cytochrome c Oxidase

For cytochrome c oxidase to function as a "molecular machine" in the way it does requires a number of metal centres. The structure of mammalian cytochrome c oxidase [79] confirms the presence of three coppers, one zinc, one magnesium and two *a*-type hemes. The positions of these centres within the enzyme are indicated schematically in Fig. 5. Cu_A is the redox centre closest to the protein surface on the side of the inter-membrane space. As such, it is ideally placed to receive electrons from cytochrome c. In the past there was some controversy concerning the number of coppers (one or two) in the Cu_A centre. Recent biochemical analysis and mutagenesis experiments supported a binuclear model [90]. The X-ray structure verifies the binuclear nature of the Cu_A site with the two coppers being 2.7 Å apart [79]. The 3-D structure also confirms that the ligands to the copper atoms are those previously predicted from mutagenesis studies [90].

Electrons are transferred from Cu_A to the first of the *a*-type hemes, which functions as a simple redox centre transferring electrons from Cu_A to the oxygen reduction site. This heme, designated as "heme-*a*", has a low-spin hexa-coordinated iron with the fifth and sixth ligands provided by the imidazole groups of two histidine residues (in the bovine enzyme these are His61 of helix II and His378 of Helix X in subunit I) [79]. It has been found that only *a*-type hemes can be extracted from cytochrome c oxidase and, since there are two heme groups in the enzyme, it is clear that any difference in the physical properties of the protein-bound hemes must arise from variations in the way in which each heme is ligated to the polypeptide. This is indeed the case and the

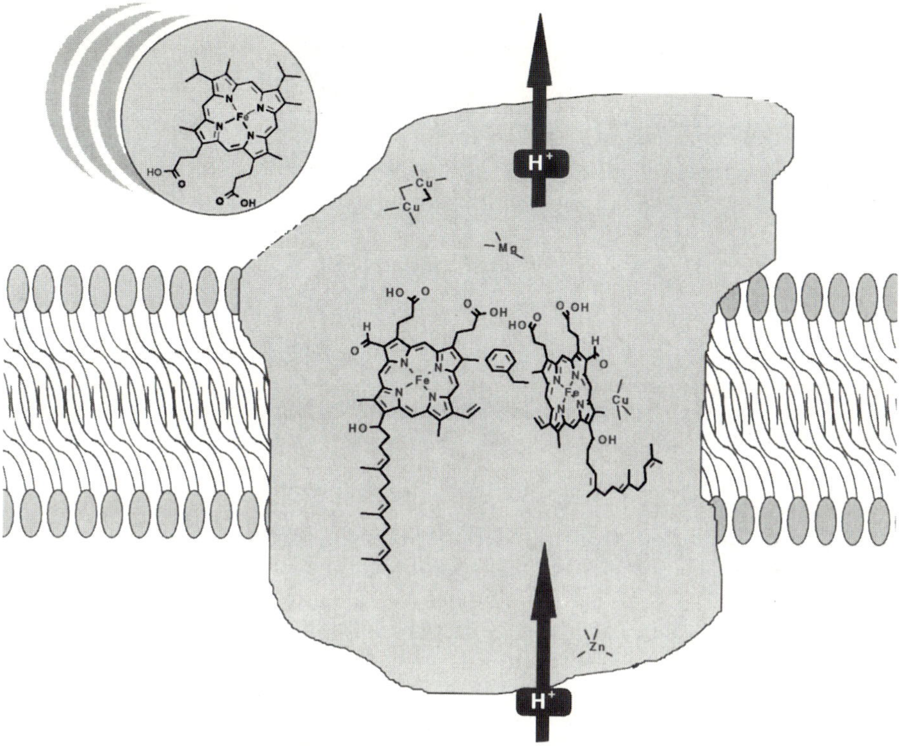

Fig. 5. A schematic representation of the crystal structure of mitochondrial cytochrome c oxidase [79]. Electrons from ferrocytochrome c are accepted by the dimeric Cu_A site before being shuttled via heme a to the heme a_3/Cu_B oxidase site. The arrival of these electrons is coupled to proton translocation across the mitochondrial inner membrane

second heme, designated as heme-a_3, has only one axial ligand from the protein (the imidazole of His376 of helix X in subunit I) [79]. The high-spin iron of heme-a_3 [91] is slightly out of the plane of the porphyrin ring pointing towards Cu_B [78, 79]. Thus, heme-a_3 together with Cu_B form a binuclear centre which is the catalytic site for oxygen reduction. Cu_B, which is coordinated by the imidazole groups of three histidine residues, is some 4.5 Å away from the iron of heme-a_3 [79] (5.2 Å in the bacterial enzyme [78]). This would allow O_2 to bind between the metal centres. The arrangement of heme-a_3 and Cu_B is illustrated in Fig. 6. The mechanism by which oxygen is reduced by cytochrome c oxidase and the nature of the intermediates involved in this reaction have been the subject of a vast number of studies which are reviewed particularly well in reference [3]. Recently, time-resolved high-resolution resonance Raman spectroscopy has proved very useful in defining the nature of some of these intermediates [92–94]. It appears that the initial intermediate is an end-on O_2 adduct of a-Fe(II) [93]. The resonance Raman data are consistent with the following intermediates being formed in the reaction:

$$a\text{-Fe}^{(II)}\,O_2 \rightarrow a\text{-Fe}^{(III)}\,\text{O-OH} \rightarrow a\text{-Fe}^{(IV)}{=}\text{O} \rightarrow a\text{-Fe}^{(III)}\,\text{OH}.$$

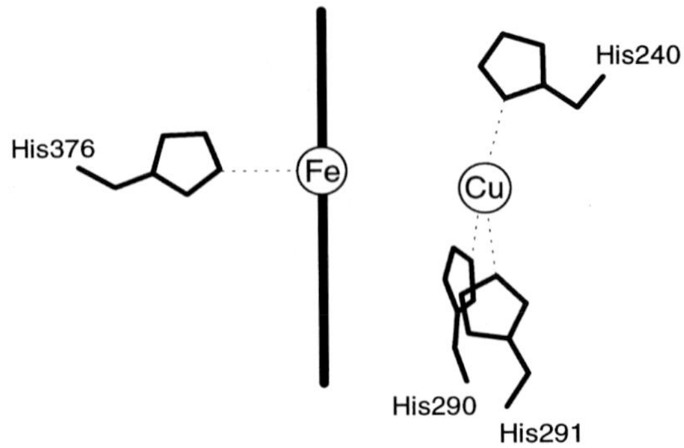

Fig. 6. A schematic representation of the heme a_3/Cu_B centre in mitochondrial cytochrome c oxidase [79]. The porphyrin ring of heme a_3, represented by a thick black line, lies perpendicular to the page

The route of the electron transfers between the prosthetic groups in cytochrome c oxidase is somewhat clarified in the light of the crystal structures [78, 79]. In particular, it is notable that hemes *a* and a_3 are almost perpendicular to each other and with an edge-to-edge distance of only about 5 Å [79]. This short distance between the heme groups is ideal for facile intra-protein electron transfer. In addition, the phenyl ring of Phe377 of subunit I in the bovine enzyme is in van der Waals contact with both heme groups, which may allow it to act as an electron conduit between the two hemes [79].

Suggestions concerning the nature of the electron and proton transfers in cytochrome oxidase, based on the recent 3-D structures, are now being tested by a number of research groups using a combination of powerful techniques from molecular biology and modern spectroscopy.

3.3 Oxygen as a Metabolite: Cytochromes P-450

There is perhaps no better example of the exploitation of heme than that seen in the cytochromes P-450 (P-450s). These remarkable enzymes use the same basic construction and chemistry to catalyse a wide variety of transformations of an enormous library of chemicals. Among the many types of chemical reactions catalysed by the P-450s are hydroxylation, epoxidation, dehydrogenation, sulphoxidation, dehalogenation and *N*-, *S*- and *O*-dealkylation [95]. In many respects, the P-450s are the best indicators of the versatility of the heme redox centre. Simple variations in the amino acid environment around the protoheme IX (heme-b) group provide the correct reaction sites to permit this huge diversity of reaction types. While the majority of the processes catalysed by the P-450s result from an initial substrate monooxygenation reaction, the P-450s can also "turn their hand" to such reactions as reduction (of alkyl halides and nitric oxide, for example), isomerisation (of prostaglandin H2) and dehydration [96–99]. Moreover, numerous eukaryotes and bacteria respond to the presence of a P-450 substrate by inducing the production, at the gene level, of the P-450 responsible for its catabolism. These elegant response mechanisms provide organisms with the ability to cope with novel exposure to an array of xenobiotics and to changes in the concentrations of endogenous biochemicals. The mammalian microsomal P-450s are a first line defence mechanism for higher organisms, playing a vital role in the conversion of xenobiotic toxins. Many microbes have taken this a step further, utilising the P-450s in peculiar metabolic pathways for the assimilation of energy contained within unusual carbon sources [e.g. 100].

3.3.1 *Classification and Properties of Cytochromes P-450*

The cytochromes P-450 are b-type cytochromes which derive their title from the absorbance peak associated with their carbon monoxide-bound ferrous forms [101, 102]. The unusual absorbance properties derive from an iron-sulphur bond (through an absolutely conserved cysteine residue) which anchors heme into the active sites of the enzyme. While there is much variation in the amino acid sequences available for the P-450s, the known structures exhibit significant similarity – all are α-helix-rich constructions resembling triagonal prisms [103]. Much interest surrounds the P-450s due to their ability to bind and "activate" molecular oxygen, breaking the dioxygen bond and catalysing the insertion of a single atom of oxygen into a wide range of organic molecules (Fig. 7), often with high degrees of stereo- and regio-specificity. A vast number

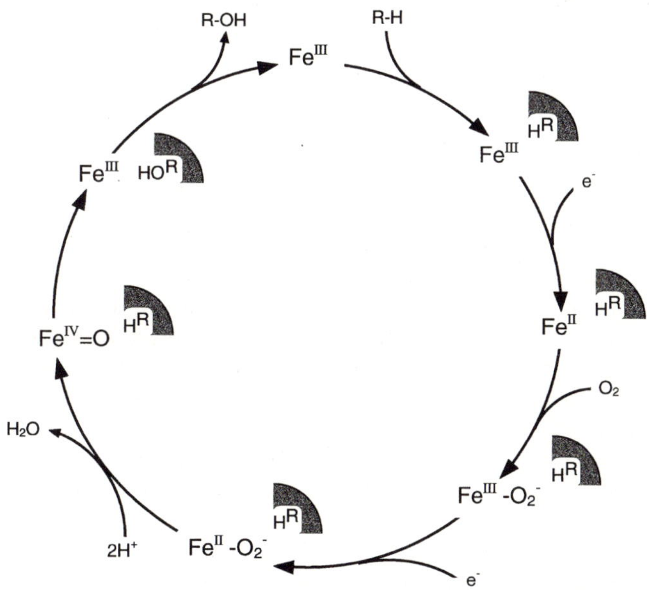

Fig. 7. A catalytic cycle showing the steps involved in a substrate hydroxylation reaction as catalysed by cytochrome P-450. Details of the intermediates are described in the main text

of P-450s have been characterised at the DNA level. Partial and total nucleotide sequences (and, hence, translated amino acid sequences) had been identified for well over 300 forms by 1993 [104]. Rather fewer have been characterised at the enzyme level – partly due to the inherent difficulties in purifying and working with membrane-bound enzymes. However, the bacterial forms of P-450 are all soluble enzymes and more amenable to purification and characterisation by protein chemical and spectroscopic techniques. It is through study of the bacterial forms (principally the camphor hydroxylase, P-450cam, from *Pseudomonas putida* and the fatty acid hydroxylase, P-450 BM3, from *Bacillus megaterium*) that most of our comprehension of P-450 structure/function has been gleaned. The three-dimensional structures of four cytochromes P-450 have been solved to date – all from bacterial sources [105–108].

The cytochromes P-450 fall into two broad classes depending on the redox partners which provide them with the two successive one-electron transfers required for the catalytic process. Class I enzymes are the terminal oxidases of a 3-component electron-transfer chain and are typified by the mammalian adrenal mitochondrial steroid hydroxylases and all but one bacterial form [109]. The P-450 is reduced by a small iron-sulphur protein (ferredoxin) which, in turn, is reduced by an NAD(P)H-dependent FAD-containing ferredoxin reductase [110]. Class II P-450s are typified by the mammalian hepatic drug-metabolising forms. In these cases, a single FAD- and FMN-containing enzyme (an NADPH-cytochrome P-450 reductase) catalyses electron transfers to the P-450 [111]. The order of electron flow is NADPH $\rightarrow$ FAD $\rightarrow$ FMN $\rightarrow$ heme. The sole bacterial class II P-450 is a particularly interesting form. Not only is P-450 BM3 from *Bacillus megaterium* ATCC 14581 reduced by a eukaryotic-like P-450 reductase, but the reductase is fused to the P-450 in a single polypeptide [112]. In eukaryotic microsomal systems, the ability of cytochrome b_5 to donate electrons to cytochrome P-450 has also been noted [113]. Cytochrome b_5 is reduced by its own NADH-dependent reductase and, hence, synergy in P-450-mediated catalysis has been shown in the presence of both NADH and NADPH.

Another convenient method of classification of the P-450s is the solubility of the various forms. Without exception, the prokaryotic members of the P-450 "superfamily" are soluble and can be maintained in aqueous solution at high concentrations. Eukaryotic forms are generally membranous – usually associated with cell membranes via hydrophobic anchors at the *N*-terminus of the enzymes. However, it has been reported that soluble eukaryotic forms can be found in yeast [114] and in plants [115]. Recently, elegant NMR studies with P-450 BM3 have demonstrated that substrate (lauric acid) moves closer to the heme centre following one-electron reduction of the iron to the ferrous form [116]. This movement brings the fatty acid sufficiently close for hydroxylation to occur at favoured positions (n-1 to n-3) on the substrate subsequent to oxygen binding and the transfer of a second electron to the catalytic centre. Substrate movement of this type does not occur in the P-450cam system and it may be the case that this phenomenon represents another convenient schism between two evolutionarily distinct forms of P-450.

Enormous research efforts have been applied to the study of human and other mammalian P-450 forms. The liver, kidney, lungs and steroidogenic organs are all rich in P-450s, and the brain is also known to contain P-450 [117]. Research has focused on a variety of aspects – all with strong medical implications. For instance, many prescribed drugs are metabolised by the hepatic P-450 systems and may also act as inducers of the production of the relevant isoforms (usually at the gene level). Barbiturates are a classic example of the P-450 induction phenomenon [118]. Analysis of the P-450s and their substrate specificity profile is central to the comprehension of the efficacy and safety of numerous drugs. The basic roles of the P-450s in human metabolism are still being deciphered and the means by which the P-450s are regulated (at both the gene and protein level) is of major importance. A very large number of human P-450s are involved in steroid metabolism and subject to regulation by hormones. Indeed, it is one of the earliest studied P-450s (the adrenal gland mitochondrial P-450scc) which catalyses the oxidative cleavage of the cholesterol side chain to produce pregnenolone – the initial step in the mammalian steroid synthesis pathway [119]. The study of P-450s involved in hormonal metabolism is a vast field with relevance for sexual development and disease states such as diabetes and liver cirrhosis [120].

Since the discovery of the first P-450s in the late 1950s in mammalian hepatic microsomal samples, there has been an explosion in the number of P-450s found. Multiple membrane-bound forms have been identified from eukaryotic sources with a plethora of important biological functions, including human liver isoforms which play a role in the oxidation of numerous drugs (e.g. ibuprofen, codeine, caffeine and haloperidol) and in the metabolism and interconversion of steroids [121]. A dramatic increase in the exposure of mammals to drugs and other environmental xenobiotics has occurred over approximately the last 50 years, and the P-450s are considered to play an important protective role against these compounds. Taken together, the multiple forms of mammalian P-450s have a wide and often overlapping substrate specificity profile. The oxidation of the xenobiotics that the P-450s catalyse (most often hydroxylations) generally result in an increase in the solubility of these compounds and facilitate their excretion in aqueous solutions of urine and/or bile. However, the P-450s can act as a "double-edged sword" – transforming otherwise inert chemicals into highly reactive and possibly mutagenic/carcinogenic derivatives. An example is the biotransformation of benzo-a-pyrene, a relatively inert component of cigarette smoke, into a carcinogenic epoxide derivative [122].

The advent of gene sequencing methodologies around 20 years ago precipitated the accumulation and comparison of translated P-450 amino acid sequences. Even by the mid-1980s (with only a few sequences available) it became clear that there was sufficient similarity between different P-450s to permit the identification of closely related forms and to derive evolutionary relationships. In recent years, improved techniques for gene isolation/characterisation and genome sequencing projects have provided a huge databank of P-450 sequences for analysis. The P-450s have been organised into an enzyme "superfamily" with divisions into families based on protein sequence homology. Within the superfamily, members of the same family generally share $\geq 40\%$ identity with one another – and members of the same sub-family show even stronger identity. Not surprisingly, P-450s with similar substrate preferences are frequently clustered in the same family. For instance, fatty acid oxygenases in family 4 and steroidogenic P-450s in family 11. In 1995, the landmark of 450 cytochrome P-450s (including genes, pseudogenes and expressed sequence tags) was passed [123] and this number now exceeds 500 [124]. Clearly, a great deal of protein chemical, structural and enzymological characterisation remains to be performed with the vast majority of the P-450 enzymes. At least forty five P-450 genes have been identified in the rat and thirty four in humans. The *C. elegans* genome may turn out to have a similar or even greater number [125]. The vast number of P-450s in certain species is an indicator of the versatility of their chemistry and of their importance to the survival of higher organisms. Analysis of P-450 evolution using amino acid sequence alignments suggests that fatty acid metabolising forms were among the earliest to develop. The fact that P-450 has been discovered in archaebacteria, ancient organisms, suggests that P-450s may have served useful roles even in very early life forms.

3.3.2
Microbial Model Systems and the P-450 Catalytic Process

The best characterised of all the P-450s is the P-450cam hydroxylase system from *Pseudomonas putida* ATCC 17453. The enzyme is one of a series encoded by genes carried on a large (230 kilobases) plasmid (the CAM plasmid) which confer on *P. putida* the ability to grow on camphor as a sole carbon source. P-450cam (encoded by the *cam* C gene) catalyses the initial step in the pathway for degradation of camphor to acetate and isobutyrate, the 5-exo hydroxylation of the monoterpene D-camphor, generating metabolic energy [126]. The P-450cam protein was purified to homogeneity almost 20 years ago [127]. Its redox partners (the flavoprotein putidaredoxin reductase and the small iron-sulphur protein putidaredoxin – encoded by the genes *camA* and *camB*, respectively) have also been purified. Genes for the 3-component hydroxylase system are transcribed as an operon, along with the product of *camD* – 5-exo-hydroxycamphor reductase. The *camDCAB* operon encodes all the genes responsible for early steps in camphor degradation. The genes for all components of the P-450cam system have been cloned and overexpressed in *E. coli*, enabling purification and *in vitro* NADH-dependent reconstitution of camphor hydroxylase function [128, 129]. The atomic structure of P-450cam was determined in 1985, and a higher resolution camphor-bound structure two years later [105]. Further strides in the analysis of the P-450 active site have been made in the laboratory of Tom Poulos at UCI, with the solution of a variety of P-450cam structures in the presence of various substrates and inhibitors [130 and references therein].

The chemistry of oxygen activation catalysed by the P-450s has always been a topic of great intrigue. The process is perhaps best described using P-450cam as the model. The heme iron of P-450 is ligated by a cysteine as the fifth ligand. The sixth position is left vacant for ligation of oxygen during catalysis – but may be occupied by a water molecule in the resting state. Substrate binding to the P-450 provides the starting point for the monooxygenase cycle (Fig. 7). The heme iron is ferric and low-spin ($S = 1/2$) until camphor binds. Thereafter, active- site water is displaced by the substrate and the iron becomes high-spin ($S=5/2$). Associated with the spin-state shift is an increase in the redox potential of the heme (from approx. -300m V to -170m V). The redox potential shift permits one-electron reduction by reduced putidaredoxin [131]. Oxygen then binds to the ferrous heme iron, forming a ferric-superoxide species referred to as "oxy-P-450" [132]. The delivery of a second electron to the heme (also from putidaredoxin) precipitates an extremely rapid substrate oxygenation process, the mechanism of which is still the subject of some controversy. A likely mechanism involves reduction to a ferric-peroxy species, which may also exist in a ferrous-superoxide form. The subsequent delivery of two protons to the reaction centre results in the loss of one atom of oxygen in the formation of a water molecule and the production of a transient, high-valent, iron-oxo species. The abstraction of a hydrogen from the camphor results in formation of a substrate radical that collapses via an oxygen rebound to generate a ferric enzyme-product complex [133, 134]. Dissociation of the 5-OH camphor returns the P-450 to its resting state. The ability of the P-450s to catalyse the oxidation of both NAD(P)H and substrate explains their older title of "mixed function oxidases". Fig. 7 depicts the various intermediate states for the heme active site seen in the catalytic cycle of a typical P450.

While P-450cam remains the most intensively studied of all the P-450s, there has also been intensive research on the P-450 BM3 system over the last decade, with the realisation that it provides an experimentally tractable model for the entire class II P-450 electron transfer chain (di-flavin P-450 reductase and P-450) in a single polypeptide. At 119,000 Da, the fatty acid hydroxylase P-450 BM3 is the largest P-450. It also has the highest monooxygenase activity of any P-450 – with rates of up to 4600 min^{-1} reported for the oxidation of favoured substrates. While P-450 BM3 will hydroxylate fatty acids with a variety of chain lengths, optimal activity is towards C15 and C16 molecules [135 and references therein]. Oxidation takes place near the terminal of the molecules. The n-2 position is usually favoured, but oxidations can also occur at n-1 and n-3 [136]. Successive rounds of oxidation (at different positions) have been shown with palmitate [137]. The sub-gene encoding the P-450 "domain" of P-450 BM3 has been cloned and overexpressed in isolation in *E. coli*, enabling purification of large quantities of the P-450 domain [138].

The crystal structure of the hemoprotein domain of P-450 BM3 was solved recently [106] (Fig. 8). Like P-450cam, the structure approximates a triangular prism; though in P-450 BM3 the active site cleft is much larger – a long hydrophobic channel of sufficient size to accommodate fatty acids of various lengths. Co-binding of the inhibitor molecules pyridine or metyrapone, simultaneously with fatty acids, can also occur in the P-450 BM3 active site [139, 140].

While site-directed mutagenesis studies on the P-450s are still in their relatively early stages, the known P-450 crystal structures provide important modelling templates for such research – allowing the identification of target residues with potential roles in redox partner docking, substrate binding and heme ligation, for example. Studies on P-450 BM3 and P-450cam have confirmed (1) the roles of the conserved cysteine in heme ligation [126]; (2) the

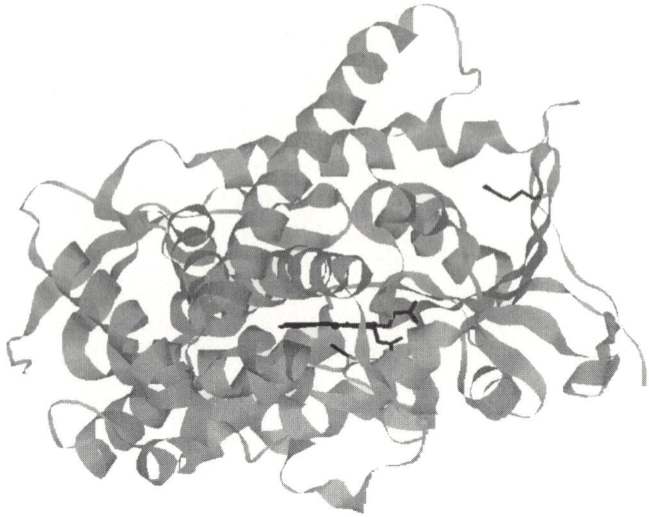

Fig. 8. The heme domain of cytochrome P-450 BM3 from the X-ray crystal structure [106]. The heme is highlighted along with the axial ligand Cys400 and a substrate binding residue Arg47

cationic residues involved in putidaredoxin binding to P-450cam [141]; (3) the residues which stabilise heme in both enzymes [142, 143]; and (4) the phenylalanine essential to substrate specificity and coupling in P-450 BM3 (Fig. 9) [144].

Future prospects for the study of the P-450s are exciting. Already, mammalian cell lines expressing different P-450 isoforms are being developed as an advanced means of screening novel drugs. While overexpression of membranous proteins is notoriously difficult, great successes have been made for a number of mammalian P-450s – by intelligently redesigning their *N*-terminal amino acids [145]. Biotechnological interest is high in the use of bacterial forms for degradation of organic pollutants and there has already been major investment in yeast P-450s for the syntheses of steroids. Recently, it has also been shown that electric currents can be used to stimulate P-450 enzymes in a productive catalytic cycle [146]. With their wealth of roles and applications in biological and chemical fields, the P-450s are certain to remain a major focus of interest in academic and pharmaceutical research institutions well into the next century.

3.4 Using Peroxides: Cytochrome c Peroxidase

Peroxides are recognised as extremely important molecules in biology. The toxic nature of hydrogen peroxide has long been known and it is clear that

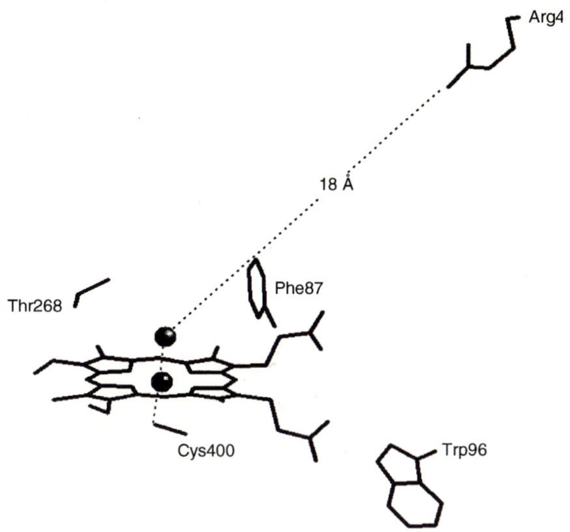

Fig. 9. The active site of cyochrome P-450-BM3 from the X-ray crystal structure [106]. The fatty-acid substrate binds via its carboxyl to Arg47 and is hydroxylated by the activated heme-oxo species near the end of its hydrophobic chain. The position of the hydroxyl ligand to Fe^{III} in this structure (shown as a grey sphere) indicates the location of this reaction. Also shown are residues: Phe87, implicated in substrate specificity and regioselectivity [144]; Trp96, a likely H-bond donor to a heme propionate (mutations at this position affect heme stability and spin-state [143]); and Cys400 which is the axial ligand to the heme-iron

enzymes such as catalase can provide a detoxification role in the cell. In spite of these possible deleterious effects it is also apparent that peroxide production and metabolism is significant in both biosynthesis and in the protection of an organism from attack.

3.4.1
Catalases and Peroxidases

In general, the catalases and peroxidases contain heme as a prosthetic group (notable exceptions include the vanadium-dependent haloperoxidases [147, 148]). Usually the heme in both classes of enzyme is heme-b, although variations include heme-*d*, which is found in the catalase HPII from *E. coli* [149]. Peroxidases and catalases have one significant similarity to the cytochromes P-450, namely the fact that they all go through an oxyferryl intermediate in their catalytic cycles. The reason that the catalases and peroxidases do not show P-450-like monooxygenation activity can be explained by differences in the protein structure around the heme. The P-450 enzymes have large binding cavities for organic substrates close to the heme which are completely absent in the catalases and peroxidases. Thus, in cytochrome P-450 the active oxygen species at the heme centre can react rapidly with the adjacent organic substrate, which would be impossible in the catalases and peroxidases.

The catalases and peroxidases have been extensively reviewed in the past [5, 150–153]. The preferred substrates for these enzymes are hydrogen peroxide and certain alkylhydroperoxides. Catalases, which are ubiquitous in aerobic organisms, promote the decomposition (or dismutation) of hydrogen peroxide to water and dioxygen as shown below:

$$2\,H_2O_2 \longrightarrow O_2 + 2\,H_2O.$$

The catalases are highly selective for H_2O_2 but will also utilise some alkylhydroperoxides, ROOH, with small R-groups. The three-dimensional structure of catalases from a number of sources are now available, including bovine liver [154], *Penicillium vitale* [155] and *Escherichia coli* [149]. The three dimensional structure of bovine liver catalase clearly shows a tyrosyl ligand proximal to the iron of the heme-b with a vacant coordination site available to bind peroxide. The heme prosthetic group is accessible to substrate via a 30 Å long channel [154]. Catalases are extremely efficient enzymes with turnovers some thousands of times higher than those of the peroxidases. They do, however, have similar properties to the peroxidases in their reaction chemistry and share a similar catalytic mechanism. The general catalytic mechanism for the catalase/peroxidase group will be illustrated in Sect. 3.4.2., entirely in the context of cytochrome c peroxidase.

The peroxidases catalyse reactions of the type shown below:

$$ROOH + XH_2 \longrightarrow ROH + H_2O + X.$$

(Where XH_2 represents a variety of reducing substrates depending on enzyme type, including phenols, amines, hydroquinones and cytochromes). There has been extensive work on peroxidases from a range of organisms including plants [156], fungi [157], yeast [158], bacteria [159] and mammals (e.g. lactoperoxidase [160], thyroid peroxidase [161] and myeloperoxidase [162]). The three-dimensional structures of a number of peroxidases are known, including Peanut peroxidase [163], chloroperoxidase [164], pea ascorbate peroxidase [165], manganese peroxidase [166], *Anthromyces ramosus* peroxidase [167], lignin peroxidase [157], and the cytochrome c peroxidases from yeast [168] and *Pseudomonas aeruginosa* [159]. To illustrate the reaction chemistry of the peroxidases (and catalases) and the intermediates formed during catalysis, we shall focus on cytochome c peroxidase from yeast.

3.4.1
Yeast Cytochrome c Peroxidase

Cytochrome c peroxidase from yeast mitochondria is a single-subunit protein with an M_r of 31 kDa (294amino acids). The enzyme catalyses the reduction of hydrogen peroxide or alkylperoxides by cytochrome c as indicated below:

$$ROOH + 2H^+ + 2\,\text{Cyt c}^{2+} \xrightarrow{\text{cytochrome c peroxidase}} ROH + H_2O + 2\,\text{Cyt c}^{3+}.$$

(Where: Cyt c^{2+} = reduced cytochrome c; Cyt c^{3+} = oxidised cytochrome c; and R = hydrogen or an alkyl group).

The three-dimensional structure of yeast cytochrome c peroxidase has been determined to 1.7 Å resolution [168] and the active site for peroxide reduction is shown in Fig. 10. The heme group is fairly well buried in the protein and the heme-iron, as shown in Fig. 10, is penta-coordinated with the imidazole group of histidine 175 providing the axial ligand. The vacant coordination site provides the locus for peroxide reduction and the side-chains of histidine 175 and arginine 48 act as distal bases, facilitating the heterolytic cleavage of the peroxide bond. Also shown in Fig. 10 , close to the heme itself, is tryptophan 191. It is the indole ring of tryptophan191 which forms a radical intermediate during the catalytic cycle [169]. The mechanism by which peroxides are reduced by cytochrome c peroxidase is essentially the same as that for all peroxidases and catalases, and proceeds as follows:

The enzyme starts from the resting state in which the heme iron is oxidised. The addition of peroxide leads to the two-electron oxidation of this resting form to produce "compound I" which is an important intermediate in all peroxidases. This step is shown below:

$$\underset{\text{resting state}}{\text{CytCP}\{\text{Fe(III)} \ldots \text{Trp191}\}} + \text{ROOH} \rightarrow \underset{\text{compound I}}{\text{CytCP}\{\text{Fe(IV)} = \text{O} \ldots \text{Trp191}^{\bullet +}\}} + \text{ROH}$$

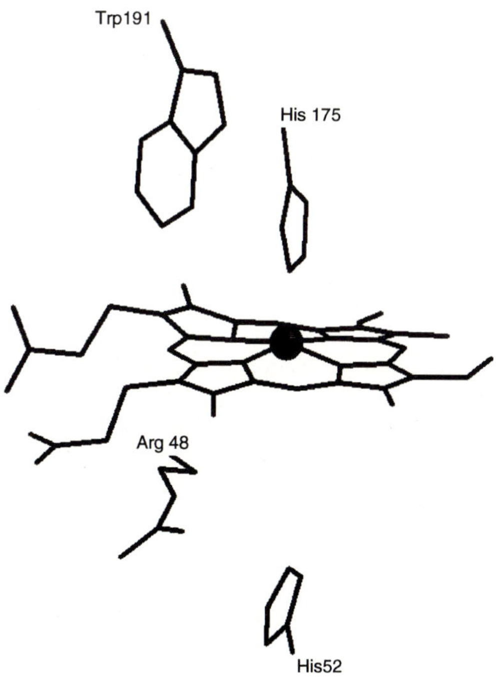

Fig. 10. The active site of yeast cytochrome c peroxidase from the X-ray crystal structure [168]. Tryptophan 191 which forms a radical intermediate during the catalytic cycle is shown, as is the axial ligand histidine 52 and the distal bases arginine 48 and histidine 52

(Where: CytCP=cytochrome c peroxidase, Fe(IV) = O = oxyferryl heme, and Trp191$^{\bullet+}$ = the radical cation form of tryptophan 191).

So, in cytochrome c peroxidase, compound I consists of an oxyferryl heme and a radical cation located on tryptophan 191. This is similar to other peroxidases, except that the π-cation radical is more typically located on the porphyrin ring [152]. To get back to the resting state requires two sequential one-electron transfers via another intermediate known as compound II [169]. These reactions are shown below:

$$\text{CytCP}\{\text{Fe(IV)} = \text{O} \ldots \text{Trp191}^{\bullet+}\} + \text{Cyt c}^{2+} \rightarrow \text{compound II} + \text{Cyt c}^{3+}$$
$$\text{compound II} + \text{Cyt c}^{2+} \rightarrow \text{CytCP}\{\text{Fe(III)} \ldots \text{Trp191}\}.$$

Compound II can exist in two forms (at least) depending on conditions (e.g. pH, ionic strength, etc) and these are either CytCP{Fe(I V) = O ... Trp191} or CytCP{Fe(III) ... Trp191$^{\bullet+}$} [169–170]. Once back at the resting state, the enzyme is able to enter another cycle of peroxide reduction.

In addition to the many studies on the mechanism of peroxide reduction, much work has also been carried out on the nature of the electron transfer from cytochrome c to cytochrome c peroxidase. The reaction between these two proteins is ideal for analysing inter-protein electron transfer (and this has been discussed at some length in Sect. 2.2.1). This is due to the fact that the crystal structures of both partners are known [168, 171] and they are both easily handled in solution. Furthermore, the three-dimensional structures of the complexes of cytochrome c peroxidase with both yeast ISO-1 cytochrome c and horse heart cytochrome c are now published [36]. Examination of these structures indicates that most of the interactions in the complexes are hydrophobic in nature [36]. This is contrary to the interactions suggested in a hypothetical model complex between the two proteins, which was essentially based on electrostatics [172]. The structures of the complexes have allowed a possible σ-tunnelling pathway to be proposed and this has already been illustrated in Fig. 3. This pathway links the heme groups of each protein via Trp191, Gly192, Ala193 and Ala194 of cytochrome c peroxidase. Thus tryptophan 191 appears to have a central role in both the formation of compound I, where the indole ring forms a radical cation, and in the electron transfer from the cytochrome c heme to the cytochrome c peroxidase heme [36] Recent site-directed mutagenesis experiments have added support to the importance of tryptophan 191 in mediating electron transfer from cytochrome c [39].

4 Production of Cellular Signals: Nitric Oxide Synthase

The nitric oxide synthases are P-450-like flavocytochromes which catalyse the production of nitric oxide (NO) from arginine and have enjoyed great scientific attention with the recent increase in understanding of the important cellular signalling roles of nitric oxide. Like P-450 BM3, the NO synthases are thiolate-ligated b-type hemoproteins fused to their FAD- and FMN-containing reductase partner in single polypeptides [173]. Reducing equivalents from NADPH drive the reaction, but there are interesting differences between the NO synthases and P-450s in their reaction and regulatory mechanisms.

Three distinct forms of mammalian NO synthase (NOS) have been recognised. The species originally found in neuronal tissue (nNOS) and endothelial cells (eNOS) are expressed constitutively and their activity is calcium ion- and calmodulin-dependent. Calmodulin binding is extremely tight [174]. In contrast, the form originally identified in macrophage cells (iNOS) can be induced by cytokines and bacterial products, and is not regulated by calmodulin or calcium. In addition, the nNOS and iNOS forms are soluble, whereas the eNOS form is membrane-bound [175]. Tetrahydrobiopterin is an important cofactor for all forms. Unlike P-450 BM3 (and the majority of the P-450s), the heme iron in NOS forms is predominantly high-spin in the resting state. Another important difference to the P-450s is that NOSs catalyse a process which requires two sequential monooxygenations. The first is the conversion of arginine to *N*-hydroxyarginine – which consumes two electrons from NADPH and produces one molecule of water. The second process requires only one electron (derived from a second NADPH molecule) and is the oxidation of *N*-hydroxyarginine to citrulline and nitric oxide. A second molecule of water is produced at this stage [176]. Interestingly, NOSs appear subject to inhibition (through heme ligation) by the NO they produce [177]

Nitric oxide is implicated in a number of important cellular functions – including control of neural transmission, cytotoxic mediation in the immune system and control of blood flow and pressure [178]. As such, the NOS systems responsible for its production have come under intense scrutiny. Enormous advances in the biochemistry and molecular biology of the NOSs have been made in the last five years – guided to a significant degree by the extensive work on the related P-450 systems which has taken place over the last quarter of a century. Genes encoding all NOS isoforms have been cloned and overexpressed, permitting biophysical studies. As with P-450 BM3, clones of sub-genes encoding the individual heme and flavin domains of a rat nNOS enzyme have been overexpressed in *E. coli* and purified. Atomic structures have yet to be obtained for the NOS forms (or their component domains) and this is certainly one focus of intense scientific competition. Other important areas include the study of the regulation of the NOS isoforms at the DNA and post-translational levels, and of their roles in relation to disease states – which may include arthritis, asthma and epilepsy [174].

5
Conclusions and the Future

It should be clear from the wealth of information reviewed in this article that the question raised in the introduction has been answered – yes, heme really is a versatile redox cofactor in biology. There can be no doubt that our knowledge of the role of heme in naturally occurring hemoproteins will continue to expand and this will be facilitated by the powerful methods in molecular and structural biology and chemistry. In addition to this, there is also the exciting new approach of designing and synthesising completely new hemoproteins based on the minimal requirements for particular functions. The possibilities of such an approach are superbly demonstrated by Dutton and colleagues who have already designed and synthesised simplified versions of multi-heme-containing proteins [179, 180].

6 References

1. Moore GR (1996) Hemeoproteins. In: Bendall DS (ed) Protein electron transfer. Bios Scientific Publishers, Oxford, p 189
2. Jameson GB and Ibers JA (1994) Biological and synthetic dioxygen carriers. In: Bertini I, Gray HB, Lippard SJ and Valentine JS (eds) Bioinorganic chemistry. University Science Books, Mill Valley CA, p 253
3. Malatesta F, Antonini G, Sarti P and Brunori M (1995) Biophys Chem 54:1
4. Ortiz de Montellano PR (1995) Cytochrome P450, 2nd edn. Plenum, New York
5. Frew J.E. and Jones P (1984) Structure and functional properties of peroxidases and catalases. In: Sykes AG (ed) Advances in inorganic and bioinorganic mechanisms, vol 3. Academic Press, London, p 175
6. Chapman SK, White SA and Reid GA (1991) Flavocytochrome b_2. In: Sykes AG (ed) Advances in Inorganic Chemistry, vol 36. Academic Press, San Diego, p 257
7. Ida S (1988) Nitrate reductase and nitrite reductase. In: Otsuka S and Yamanaka T (eds) Metalloproteins: Chemical properties and biological effects. Elsevier, Amsterdam, p 406
8. Scott RA and Mauk AG (1996) Cytochrome c: A multidisciplinary approach. University Science Books, Sausalito
9. Pettigrew GW and Moore GR (1987) Cytochromes c: Biological aspects. Springer-Verlag, Berlin Heidelberg New York
10. Moore GR and Pettigrew GW (1990) Cytochromes c: Evolutionary, structural and physicochemical aspects. Springer-Verlag, Berlin Heidelberg New York
11. Silvestrini MC, Falcinelli S, Ciabatti I, Cutruzzolà F and Brunori M (1994) Biochimie 76:641
12. Walsh TA, Johnson MK, Barber D, Thomson AJ, Greenwood C (1980) J Inorg Biochem. 14:15
13. Marcus RA (1956) J Chem Phys 24:966
14. Marcus RA (1965) J Chem Phys 43:679
15. Marcus RA (1964) Ann Rev Phys Chem 15:155
16. Marcus RA, Sutin N (1985) Biochim Biophys Acta 811:265
17. Moser CC, Keske JM, Warncke K, Farid RS, Dutton PL (1992) Nature 355:796
18. Farid RS, Moser CC, Dutton PL (1993) Curr Opinion Struct Biol 3:225
19. Chapman SK, Mount AR (1995) Nat Prod Reps 12:93
20. Beratan DN, Onuchic JN, Gray HB (1991) In: Sigel H, Sigel A (eds) Metal ions in biological systems vol 27. Marcel Dekker, New York, p 97
21. Canters GW, van de Kamp M (1992) Curr Opinion Struct Biol 2:859
22. Bendall DS (ed) (1996) Protein electron transfer. Bios Scientific Publishers, Oxford
23. Beratan DN, Onuchic JN (1996) The protein bridge between redox centres. In: Bendall DS (ed) Protein electron transfer. Bios Scientific Publishers, Oxford, p 23
24. Kassner RJ (1972) Proc Natl Acad Sci USA 69:2263
25. Raphael AC, Gray HB (1989) Proteins Struct Funct Genet 6:338
26. Campos AP, Aguiar AP, Hervas M, Regalla M, Navarro JA, Ortega JM, Xavier AV, De la Rosa MA, Teixeira M (1993) Eur J Biochem 216:329
27. Bertrand P, Bruschi M, Denis M, Gayda JP, Manca F (1982) Biochem Biophys Res Comm 106:756
28. Holton RW, Myers T (1967) Biochim Biophys Acta 131:362
29. Krogmann DW (1991) Biochim Biophys. Acta 1058:35
30. Mus-Veteau I, Dolla A, Guerlesquin F, Payan F, Czjzek M, Haser R, Bianco P, Haladjian J, Rapp-Giles B, Wall JD, Voordouw G, Bruschi M (1992) J Biol Chem 267:16851

31. Gunner MR, Honig B (1991) Proc Natl Acad Sci USA 88:9151
32. Watt GD, Frankel RB, Papaefthymiou GC, Spartalian K, Stiefel EI (1986) Biochemistry 25:4330
33. Fan K, Akutsu H, Kyogoku Y, Niki K (1990) Biochemistry 29:2257
34. Santos H, Moura JJG, Moura I, Legall J, Xavier AV (1984) Eur J Biochem 141:283
35. Mathews FS, Durley RCE (1996) Structure of electron transfer proteins and their complexes In: Bendall, DS (ed) Protein electron transfer. Bios Scientific Publishers, Oxford, p99
36. Pelletier H, Kraut J (1992) Science 258:1748
37. Millet F, Miller MA, Geren L, Durham B (1995) Bioenerg Biomemb 27:341
38. Pappa HS, Poulos TL (1995) Biochemistry 34:6573
39. Pappa HS, Tajbaksh S, Saunders AJ, Pielak GJ, Poulos TL (1996) Biochemistry 35:4837
40. Zhou JS, Hoffmann BM (1994) Science 265:1693
41. Zhou JS, Nocek JM, Devan ML, Hoffmann BM (1995) Science 269:204
42. Mauk MR, Ferrer JC, Mauk AG (1994) Biochemistry 33:12609
43. Chitnis PR, Xu Q, Chitnis VP, Nechustai R (1995) Photosynth Res 44:23
44. Hill KL, Merchant S (1995) EMBO J 14:857
45. Hervás M, Navarro JA, Díaz A, Botten H, Dela Rosa M (1995) Biochemistry 34:11321
46. Hervás M, Navarro JA, Díaz A, Dela Rosa M (1996) Biochemistry 35:2693
47. Redinbo MR, Yeates TO, Merchant S (1994) J Bioenerg Biomemb 26:49
48. Kerfield CA, Anwar HP, Interante R, Merchant S, Yeates TO (1995) J Mol Biol 250:627
49. Frazao C, Soares CM, Carrondo MA, Pohl E, Dauter Z, Wilson KS, Hervas M, Navarro JA, De la Rosa MA, Sheldrick GM (1995) Structure 3:1159
50. Banci L, Bertini I, Quacquarini G, Walter O, Diaz A, Hervas M, De la Rosa MA (1996) J Biol Inorg Chem 1:330
51. Chen L, Durley RCE, Mathews FS, Davidson VL (1994) Science 264:86
52. Lederer F (1991) Flavocytochrome b_2. In: Müller F (ed) Chemistry and biochemistry of flavoenzymes vol 2. CRC, Boca Raton FL p153
53. Pajot P, Claisse ML (1974) Eur J Biochem 49:275
54. Balme A, Brunt CE, Pallister RL, Chapman SK, Reid GA (1995) Biochem J 309:601
55. Sharp RE, White P, Chapman SK, Reid GA (1994) Biochemistry 33:5115
56. Sharp RE, Chapman SK, Reid GA (1996) Biochemistry 35:891
57. Sharp RE, Chapman SK, Reid GA (1996) Biochem J 316:507
58. Xia Z.-X, Mathews FS (1990) J Mol Biol 212:837
59. Miles CS, Rouvière-Fourmy N, Lederer F, Mathews FS, Reid GA, Black MT, Chapman SK (1992) Biochem J 285:187
60. Daff S, Ingledew WJ, Reid GA, Chapman SK (1996) Biochemistry 35:6345
61. Daff S, Sharp RE, Short DM, Bell C, White P, Manson FDC, Reid GA, Chapman SK (1996) Biochemistry 35:6351
62. Moser CC, Page CC, Farid R, Dutton PL (1996) J Bioenerg Biomemb 27:263
63. Morris CJ, Black AC, Pealing SL Manson FDC, Chapman SK, Reid GA (1994) Biochem J 302:587
64. Pealing SL, Black AC, Manson FDC, Ward FB, Chapman SK, Reid GA (1992) Biochemistry 31:12132
65. Pealing SL, Cheesman MR, Reid GA, Thomson AJ, Ward FB, Chapman SK (1995) Biochemistry 34:6153
66. Cusanovich MA, Hazzard JH & Wilson GS (1994) Advances in chemistry series 235, 471
67. Fulop V, Moir JWB, Ferguson SJ, Hajdu J (1995) Cell 81:369
68. Silvestrini MC, Falcinelli S, Ciabatti I, Cutruzzola F, Brunori M (1994) Biochimie 76: 641
69. Silvestrini MC, Cutruzzola F, Schinina ME, Maras B, Rolli G, Brunori M (1996) J Inorg Biochem 62:77
70. Silvestrini MC, Tordi MG, Musci G, Brunori M (1990) J Biol Chem 265:11783
71. Schichman SA, Meyer TA, Gray HB (1996) Inorg Chim Acta 243:25
72. Frolow F, Kalib AJ, Yariv J (1994) Nature Struct Biol 1:453

73a. Le Brun NE, Andrews SC, Guest JR, Harrison PM, Moore GR, Thomson AJ (1995) Biochem J 312:385
73b. Andrews SC, Le Brun NE, Barynin V, Thomson AJ, Moore GR, Guest JR, Harrison PM (1995) J Biol Chem 270:23268
73c. Kadir FHA, Al-Massad FK, Moore GR (1992) Biochem J 282:867
74. Garg RP, Vargo CJ, Cui X, Kurtz DM (1996) Biochemistry 35:6297
75. Hill HAO, Tew DG (1987) Dioxygen, superoxide and peroxide. In: Wilkinson G, Gillard RD, McLeverty JA (eds) Comprehensive Coordination Chemistry, vol 2. Pergamon, Oxford, p 315
76. Yagi K (1993) Flavins and Flavoproteins. Walter de Gruyter, Berlin
77. Ostermeier C, Iwata S, Michel H (1996) Curr Opinion Struct Biol 6:460
78. Iwata S, Ostermeier C, Ludwig B, Michel H (1995) Nature 36:660
79. Tsukihara T, Aoyama H, Yamashita E, Tomizaki T, Yamaguchi H, Shinzawa-Itoh K, Nakashima R, Yaono R, Yoshikawa S (1995) Science 269:1069
80. Gennis R, Fergusson-Miller S (1995) Science 269:1063
81. Trumpower BL, Gennis RB (1994) Annu Rev Biochem 63:675
82. Wikström M (1989) Nature 338:776
83. Wikström M, Bogachev A, Finel M, Morgan JE, Puustinen A, Raito M, Verkhovskaya M, Verkhovsky MI (1994) Biochim Biophys Acta 1187:106
84. Rich PR, Electron transfer complexes coupled to ion translocation. In: Bendall, DS (ed) Protein electron transfer. Bios Scientific Publishers, Oxford, p 217
85. Babcock GT, Wikström M (1992) Nature 356:301
86. Abrahams JP, Leslie AGW, Lutter R, Walker JE (1994) Nature 370:621
87. Calhoun MW, Thomas JW, Gennis RB (1994) Trends Biochem Sci 19:325
88. Fetter JR, Qian J, Shapleigh J, Thomas JW, Garcia Horsman A, Schmidt E, Hosler J, Babcock GT, Gennis RB, Fergusson-Miller S (1995) Proc Natl Acad Sci USA 92:1604
89. Morgan JE, Verkhovsky MI, Wikström M (1994) J Bioenerg Biomembr 26:599
90. Kelly M, Lappalainen P, Talbo G, Haltia T, Vanderoost J, Saraste M (1993) J Biol Chem 268:16781
91. Calhoun MW, Thomas JW, Hill JJ, Hosler JP, Shapleigh JP, Tecklenburg MMJ, Fergusson-Miller S, Babcock GT, Alben JO, Gennis RB (1993) Biochemistry 32:10905
92. Varotsis C, Zhang Y, Appelman EH, Babcock GT (1993) Proc Natl Acad Sci USA 90:237
93. Ogura T, Takahashi S, Hirota S, Shinzawa-Itoh K, Yoshikawa S, Appelman EH, Kitigawa T (1993) J Amer Chem Soc 115:8527
94. Ogura T, Hirota S, Proshlyakov DA, Shinzawa-Itoh K, Yoshikawa S, Kitigawa T (1996) J Amer Chem Soc 118:5443
95. Groves JT, Han Y-Z (1995) Models and mechanisms of cytochrome P450 action. In: Ortiz de Montellano PR (ed) Cytochrome P450, 2nd Edition. Plenum, New York, p 3
96. Nakahara K, Tanimoto T, Hatano, K, Usuda K, Shoun H (1993) J Biol Chem 268:8350
97. Hecker M, Ullrich V (1989) J Biol Chem 264:141
98. Song WC, Brash AR (1991) Science 253:781
99. Mansuy D, Renaud J-P (1995) Heme-thiolate proteins different from cytochromes P450 catalysing monooxygenations. In: Ortiz de Montellano PR (ed) Cytochrome P450, 2nd Edition. Plenum, New York, p 537
100. Hedegaard J, Gunsalus IC (1965) J Biol Chem 240:4038
101. Garfinkel D (1958) Arch Biochem Biophys 77:493
102. Klingenberg M (1958) Arch Biochem Biophys 75:376
103. Poulos TL, Cupp-Vickery J, Li H (1995) Structural studies on prokaryotic cytochromes P450. In: Ortiz de Montellano PR (ed) Cytochrome P450, 2nd Edition. Plenum, New York, p 125
104. Nelson DR, Kamataki T, Waxman DJ, Guengerich FP, Estabrook RW, Feyereisen R, Gonzalez FJ, Coon MJ, Gunsalus IC, Gotoh O, Okuda K, Nebert DW (1993) DNA Cell Biol 12:1
105. Poulos TL, Finzel BC, Howard AJ (1987) J Mol Biol 195:687
106. Ravichandran KG, Boddupalli SS, Hasemann CA, Peterson JA, Deisenhofer J (1993) Science 261:731

107. Hasemann CA, Ravichandran KG, Peterson JA, Deisenhofer J (1994) J Mol Biol 236:1169
108. Cupp-Vickery JR, Poulos TL (1995) Nature Struct Biol 2:144
109. Munro AW, Lindsay JG (1996) Mol Microbiol 20:1115
110. Huang JJ, Kimura T (1973) Biochemistry 12:406
111 Vermillion JL, Coon MJ (1974) Biochem Biophys Res Commun 60:849
112. Narhi LO, Fulco AJ (1987) J Biol Chem 262:6683
113. Hildebrandt A, Estabrook RW (1971) Arch Biochem Biophys 143:66
114. Nakayama N, Takemae A, Shoun H (1996) J Biochem 119:435
115. Schuller MA (1996) Crit Rev Plant Sci 15:235
116. Modi S, Sutcliffe MJ, Pimrose WU, Lian LY, Roberts GCK (1996) Nature Struct Biol 3:414
117. Strobel HW, Kawashima H, Geng J, Bergh A, Hodgson AV Wang HM, Shen SJ (1995) Toxicol Lett 82–83:639
118. Whitlock Jr JP, Denison MS (1995) Induction of cytochrome P450 enzymes that metabolize xenobiotics. In: Ortiz de Montellano PR (ed) Cytochrome P450, 2nd Edition. Plenum, New York, p 367
119. Simpson ER (1979) Mol Cell Endocrinol 13:213
120. Waxman DJ, Chang TKH (1995) Hormonal reguation of liver cytochrome P450 enzymes. In: Ortiz de Montellano PR (ed) Cytochrome P450, 2nd Edition. Plenum, New York, p391
121. Guengerich FP (1995) Human cytochrome P450 enzymes in: Ortiz de Montellano PR (ed) Cytochrome P450, 2nd Edition. Plenum, New York, p 473
122. Crespi CL, Penman BW, Steimel DT, Gelboin HV, Gonzalez FJ (1991) Carcinogenesis 12: 355
123. Nelson DR (1995) Conference presentation and abstract, 3rd International Conference on Cytochrome P-450 Biodiversity. Woods Hole, Massachusets Oct 8–12
124. Nelson DR, Koymans L, Kamataki T, Stegeman JJ Feyereisen R, Waxman DJ, Sects. Waterman MR, Gotoh O, Coon MJ, Estabrook RW, Gunsalus IC And Nebert DW (1996) Pharmacogenetics 6:1
125. Nelson DR (1995) Cytochrome P450 nomenclature and alignment of selected sequences. In: Ortiz de Montellano PR (ed) Cytochrome P450, 2nd Edition. Plenum, New York, p 575
126. Martinis SA, Ropp JD, Sligar SG, Gunsalus IC (1991) In: Ruckpaul K, Rein H (eds) Microbial and plant cytochromes P-450 Frontiers in Biotransformation Vol. 4. Taylor and Francis, London, p 54
127. Gunsalus IC, Wagner GC (1978) Methods Enzymol 52:166
128. Unger BP, Gunsalus IC, Sligar SG (1986) J Biol Chem 261:1158
129. Peterson JA, Lorence MC, Amarneh B (1990) J Biol Chem 265:6066
130. Raag R, Poulos TL (1991) Biochemistry 30:674
131. Sligar SG (1976) Biochemistry 15:5399
132. Sligar SG, Lipscomb JD, Debrunner JD, Gunsalus IC (1974) Biochem Biophys Res Commun 61:290
133. Mueller J, Loida PL, Sligar SG (1995) Twenty five years of P450cam research. In: Ortiz de Montellano PR (ed) Cytochrome P450, 2nd Edition. Plenum, New York, p 83
134. White RE, Coon MJ (1980) Annu Rev Biochem 49:315
135. Narhi LO, Fulco AJ (1986) J Biol Chem 261:7160
136. Miura Y, Fulco AJ (1974) J Biol Chem 249:1880
137. Boddupalli SS, Estabrook RW, Peterson JA (1990) J Biol Chem 265:4233
138. Miles JS, Munro AW, Rospendowski BN, Smith WE, McKnight, J, Thomson AJ (1992) Biochem J 303:423
139. Modi S, Primrose WU, Boyle JMB, Gibson CF, Lian LY, Roberts GCK (1995) Biochemistry 34:8982
140. MacDonald IDG, Smith WE, Munro AW (1996) FEBS Lett 396:196
141. Stayton PS, Sligar SG (1990) Biochemistry 29:7381
142. Sligar SG, Filipovic D, Stayton PS (1991) Methods Enzymol 206:31
143. Munro AW, Malarkey K, Lindsay JG, Coggins JR, Price NC, Kelly SM, McKnight J, Thomson, AJ, Miles JS (1994) Biochem J 303, 423

144. Alworth WL, Mullin DA, Xia Q, Kang L, Liu HM, Zhao W (1995) FASEB J 9: A1491
145. Fisher CW, Caudle DL, Martin-Wixtrom C, Quattrochi LC, Turkey RH, Waterman MR, Estabrook RW (1992) FASEB J 6:759
146. Faulkner KM, Shet MS, Fisher CW, Estabrook RW (1995) Proc Natl Acad Sci USA 92: 7705
147. Weyand M, Hecht HJ, Vilter H, Schonburg D (1996) Acta Cryst D 52:864
148. Messerschmidt A, Wever R (1996) Proc Natl Acad Sci USA 93:392
149. Bravo J, Verdaguer N, Tormo J, Betzel C, Swittala J. Loewen PC, Fita I (1995) Structure 3: 491
150. Everse J, Everse KE, Grisham MB (1991) Peroxidases in Chemistry, and Biology Vol 2. CRC, Boca Raton Fl
151. Valentine JS (1994) Dioxygen reactions. In: Bertini I, Gray HB, Lippard SJ, Valentine JS (eds) Bioinorganic Chemistry. University Science Books, Mill Valley CA, p 295
152. Yamazaki I (1988) Peroxidase. In: Otsuka S, Yamanaka T (eds) Metalloproteins: Chemical properties and biological effects. Elsevier, Amsterdam, p 218
153. Yamazaki I (1988) Catalase. In: Otsuka S, Yamanaka T (eds) Metalloproteins: Chemical properties and biological effects. Elsevier, Amsterdam, p 224
154. Murphy MRN, Reid TJ, Sicignano A, Tanaka N, Rossmann MG (1981) J Mol Biol 152:465
155. Vainshtein BK, Melik-Adamyan WR, Barynin VV, Vagin AA, Grebenko AI (1981) Nature 293:411
156. Poulos TL, Kraut J (1980) J Biol Chem 225:8199
157. Edwards SL, Raag R, Wariishi H, Gold MH, Poulos TL (1993) Proc Natl Acad Sci USA 90: 750
158. Bosshard HR, Anni H, Yonetani T (1991) Yeast cytochrome c peroxidase. In: Everse J, Everse KE, Grisham MB (eds) Peroxidases in Chemistry and Biology Vol 2. CRC, Boca Raton Fl, p 51
159. Fulop V, Ridout CJ, Greenwood C, Hajdu J (1995) Structure 3:1225
160. Andersson LA, Bylhas SA, Wilson AE (1996) J Biol Chem 271:406
161. Taurog A, Davis ML, Doerge DR (1996) Arch Biochem Biophys 330:24
162. Davey CA, Fenna RE (1996) Biochemistry 35:10967
163. Schuller DJ, Ban N, Vanhystee RB, McPherson A, Poulos TL (1996) Structure 4:311
164. Sundaramoorthy M, Terner J, Poulos TL (1996) Structure 3:1367
165. Patterson WR, Poulos TL (1995) Biochemistry 34:4331
166. Sundaramoorthy M, Kishi K, Gold MH, Poulos TL (1994) J Biol Chem 269:32759
167. Kunishima N, Fukuyama K, Matsubera H, Hatanaka H, Shibaro Y, Amachi T (1994) J Mol Biol 235: 331
168. Finzel BC, Poulos TL, Kraut J (1984) J Biol Chem 21:13027
169. Bonagura CA, Sunaramoorthy M, Pappa HS, Patterson WR, Poulos TL (1996) Biochemistry 35:6107
170. Mondal MS, Fuller HA, Armstrong FA (1996) J Am Chem Soc 118:263
171. Bushnell GW, Louie GV, Brayer GD (1990) J Mol Biol 214:585
172. Poulos TL, Kraut J (1980) J Biol Chem 255:10322
173. White KA, Marletta MA (1992) Biochemistry 31:6627
174. Knowles R (1994) The Biochemist Oct/Nov 1994, 3
175. Zembowicz A, Hatchett RJ, Radziszewski W, Jakubowski AM, Gryglewski RJ (1994) In: Moncada S, Feelisch M, Busche R, Higgs EA (eds) The Biology of Nitric Oxide. 4: Enzymology, Biochemistry and Immunology. Portland Press, London, p 11
176. Prince RC, Gunson DE (1993) Trends Biochem Sci 18:35
177. Griscavage JM, Fukuto JM, Komori Y, Ignarro LJ (1994) J Biol Chem 269:21644
178. Wolff DJ, Lubeskie A, Umansky S (1994) Arch Biochem Biophys 314:360
179. Robertson DE, Farid RS, Moser CC, Urbauer JL, Mulholland SE, Pidikiti R, Lear JD, Wand AJ, DeGrado WF, Dutton PL (1994) Nature 368:425
180. Farid RS, Robertson DE, Moser CC, Pilloud D, DeGrado WF, Dutton PL (1994) Biochem Soc Trans 22:689

Rationalisation of Metal Binding to Transferrin: Prediction of Metal-Protein Stability Constants

Hongzhe Sun*, Mark C. Cox, Hongyan Li and Peter J. Sadler*

Department of Chemistry, University of Edinburgh, King's Buildings, West Mains Road, Edinburgh EH9 3JJ, UK. *E-mail: h.sun@ed.ac.uk; p.j.sadler@ed.ac.uk*

Serum transferrin is an 80 kDa glycoprotein which transports Fe^{3+} and a variety of metal ions of diagnostic, therapeutic and toxic importance. Methods for the study of the differences between the two metal binding sites (the N- and C-lobe sites), and mechanisms for the uptake and release of both metal ions and synergistic anions are discussed. The strength of metal binding to transferrin can be rationalised on the basis of metal ion acidity (strength of hydroxide binding). Similarly the strength of metal binding to the enzymes carbonic anhydrase and carboxypeptidase can be correlated with the binding of the same metal ions to imidazole. Such correlations of metal binding to proteins and low molecular mass ligands may provide insight into the preorganisation of metal binding sites in proteins.

Keywords: transferrin; carbonic anhydrase; carboxypeptidase; stability constant; metal ions.

* Correspondence to: Dr Hongzhe Sun, Professor Peter J. Sadler, Department of Chemistry, University of Edinburgh, King's Buildings, West Mains Road, Edinburgh EH9 3JJ, UK.

Abbreviations

oTF	Ovotransferrin;
oTF/2N	*N*-terminal half-molecule of chicken ovotransferrin;
oTF/2C	*C*-terminal half-molecule of chicken ovotransferrin;
EDDA	*N,N'*-ethylenediaminediacetate;
hTF	human serum transferrin;
hTF/2N	*N*-terminal half-molecule of human serum transferrin;
hLF	human lactoferrin;
HMQC	heteronuclear multiple-quantum coherence;
K^*	bicarbonate-independent binding constant;
NTA	nitrilotriacetate;
rTF	rabbit serum transferrin.

1 Introduction

Transferrins are single-chain glycoproteins containing about 700 amino acids with molecular masses of approximately 80 kDa (human serum transferrin (hTF), 679 amino acids 79,750 Da; chicken ovotransferrin (oTF), 686 amino acids, 78,000 Da). The functions of transferrins are iron transport (in blood) and antimicrobial activity (ovotransferrin and lactoferrin). Recognition of transferrin by cells is via receptor-mediated endocytosis. Only diferric transferrin (two iron sites occupied) and not the apo-form binds strongly to the receptor protein on the surface of cells and is taken up. Inside the cell, diferric transferrin is held in membrane-bound vesicles (endosomes) where the pH is lowered from the extracellular value of 7.4 to about 5.5, and Fe^{3+} is released. Apo-transferrin remains bound to its receptor due to its high affinity for the receptor at acid pH, and it is recycled back to the surface of the cell. At extracellular physiological pH, apo-transferrin dissociates from its receptor due to its low affinity at pH 7.4, is released into the circulation, and reutilised [1, 2].

Table 1. X-ray crystal structures of transferrins

Transferrin	Residues	Metal binding ligands [a]	Geometry	Resolution	pH	Comments	Refs.
Fe_c-hTF	679	C-lobe: Y426, Y517, H585, D392 and bidentate carbonate	distorted octahedral	low		N-lobe open (apo), C-lobe closed (Fe^{3+})	[3]
apo-hLF	692			2.8 Å	8.2	N-lobe open and C-lobe closed	[4]
Fe_2-rTF	676	N-lobe: Y95, Y188, H249, D63 and bidentate carbonate; C-lobe: Y426, Y517, H585, D392 and bidentate carbonate	distorted octahedral	3.3 Å	6.0		[9]
Fe_2-hLF	692	N-lobe: Y92, Y192, H253, D60 and bidentate carbonate; C-lobe: Y435, Y528, H597, D395 and bidentate carbonate.	octahedral	2.2 Å	7.8	Carbohydrate partially defined. Crystallisation conditions: 10 mM phosphate, 10% (v/v) ethanol.	[5–7]
Fe_2-oTF	686	N-lobe: Y92, Y191, H250, D60 and bidentate carbonate; C-lobe: Y431, Y524, H592, D395 and bidentate carbonate	distorted octahedral	2.4 Å	5.9	Crystallisation conditions: 100 mg/ml in 20 mM sodium acetate buffer pH 5.9, 4 °C, addition of 2 $\mu l\ ml^{-1}$ of a 20% (w/v) PEG 6000 solution	[8]
Cu_2-ox-hLF	692	N-lobe: Y92, Y192, H253, D60 and monodentate oxalate; C-lobe: Y435, Y528, H597, D395 and bidentate oxalate.	N-lobe: square pyramidal, penta coordinate; C-lobe: octahedral	2.8 Å	7.8	Oxalate as anion	[43]
Fe-hLF/2N	333	Y92, Y192, H253, D60 and bidentate carbonate	octahedral	2.0 Å	8.0		[44]

Table 1 (continued)

Transferrin	Residues	Metal binding ligands [a]	Geometry	Resolution	pH	Comments	Refs.
Fe-Asp60Ser Mutant hLF/2N	333	Y92, Y192, H253, bidentate carbonate and a water ligand	octahedral	2.05 Å	8.0	Crystallisation conditions: 60 mg/ml, 10 mM Tris-HCl, 12% (v/v) isopropanol	[18]
NII-domain fragment (Fe)-duck oTF	249	Y95, Y188, bidentate carbonate and possible glycine or two waters	octahedral	2.3 Å	7.8	Crystallisation conditions: 40 mg/ml, 50 mM-Tris buffer, 40% (v/v) 2-methyl-2,4-pentanediol, 4 °C	[46]

[a] Numbering of residues: Y92 (435) in hLF is equivalent to Y95 (426) in hTF or oTF.

The three-dimensional structures of monoferric human serum transferrin (Fe_C-hTF with Fe^{3+} in the C-lobe) [3], apo- and diferric human lactoferrin (hLF) [4–7], diferric hen ovotransferrin (hen oTF) [8] and diferric rabbit serum transferrin (rTF)[9] have been determined by X-ray crystallography (Table 1). The X-ray crystal structures show that transferrins are bilobal with two structurally similar lobes (N-lobe and C-lobe) each of 40 kDa joined by a short peptide (Fig. 1). Each lobe is further divided into two domains of similar size which

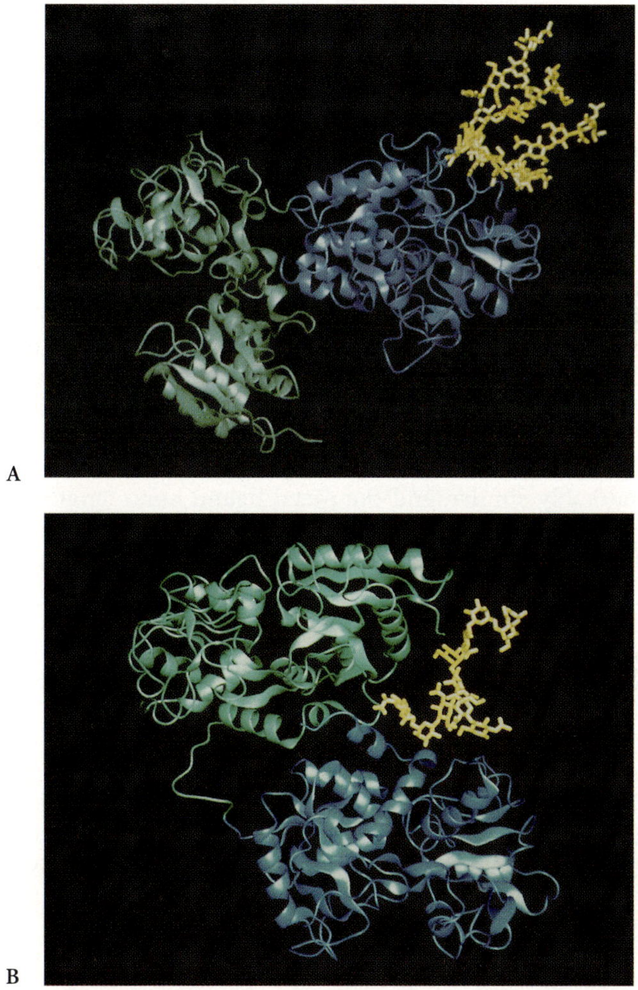

Fig. 1. A. *Top*: A model of human serum transferrin based on the X-ray crystal structure of human Fe_C-transferrin [3]. The C-lobe, which contains bound Fe^{3+}, is shown in a closed form (blue) and the apo N-lobe in an open form (green). The two biantennary glycan chains in the C-lobe attached to Asn416 and Asn611 are shown in yellow, but are not seen by crystallography and were computer-modelled. **B** *Bottom*: diferric rabbit serum transferrin (Fe_2-rTF); both lobes contain bound Fe^{3+} and are closed [based on Ref. 9]. The glycan chain at Asn490 was computer-modelled

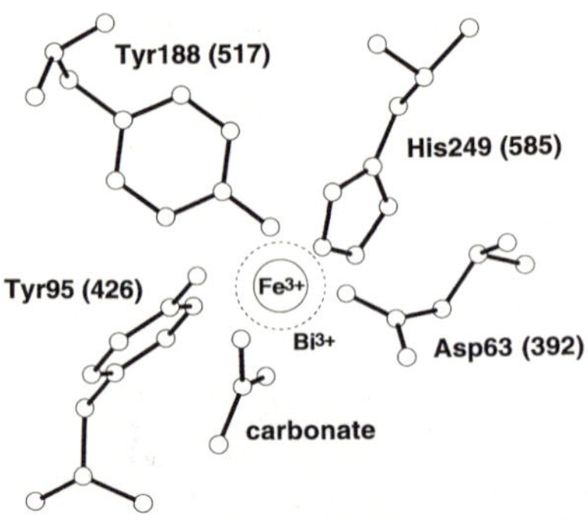

Fig. 2. A model of the N-lobe and C-lobe metal binding sites of human serum transferrin, illustrating the difference in size between Fe^{3+} and Bi^{3+} (dotted line). The residue numbers in brackets refer to the C-lobe.

have alternating α-helical and β-sheet segments (lactoferrin N-lobe: domain I residues 1–90 and 252–333, domain II residues 91–251), a common feature for binding proteins of the "Venus fly-trap" family [10]. These domains may have a crucial functional significance since the cleft separating the two domains of each lobe houses the metal binding site. Each lobe contains a distorted octahedral Fe^{3+} binding site consisting of two Tyr, one His, one Asp and one bidentate carbonate anion (the so-called synergistic anion). These two binding sites are remarkably similar, and the metal-ligand bond lengths are all about 1.9 ~ 2.2 Å. Fig. 2 shows the Fe^{3+} binding sites of human serum transferrin. The ligands are from four different parts of the protein structure, one from domain 1 (Asp63 in human transferrin), one from domain 2 (Tyr188) and two from the two polypeptide strands (Tyr95 and His 249) that cross over between the two domains at the back of the iron site [3]. Thus the domains can move apart to a more open conformation, hinged by the backbone strands. This allows release of Fe^{3+} from the site.

An intriguing aspect of the chemistry of transferrin is that Fe^{3+} cannot bind strongly without concomitant binding of a synergistic anion. The common features of the synergistic anion are a carboxylate donor and a second (proximal) electron donor two carbon atoms removed. Carbonate serves as the synergistic anion in vivo and has a higher affinity than most other anions, but oxalate is also efficient in promoting metal ion binding. Both carbonate and oxalate bind to Fe^{3+} in a bidentate mode [4–9, 11].

2
Metal-Induced Conformational Changes

Transferrin has long been known to undergo conformational changes during Fe^{3+} uptake and release. The conformational changes are likely to be of functional importance, and play a crucial role in receptor recognition. X-ray crystallographic studies of Fe_NFe_C-lactoferrin (iron in both sites) and apo-lactoferrin

Table 2. Radii of gyration for transferrins

Samples	R_g (Å) [a]	ΔR_g (Å) [a]	References
apo-hTF	32.5 ± 0.2		[14]
Fe^{3+}_2-hTF	31.4 ± 0.2	– 1.1	[14]
apo-oTF	30.5 ± 0.2		[14]
Fe^{3+}_2-oTF	29.7 ± 0.2	– 0.8	[14]
Cu^{2+}_2-oTF	29.9	– 0.6	[13]
In^{3+}_2-oTF	29.9	– 0.6	[13]
Al^{3+}_2-oTF	30.1	– 0.4	[13]
Hf^{4+}_2-oTF	30.4	– 0.1	[13]
apo-hLF	33.3 ± 0.2		[14]
Fe^{3+}_2-hLF	33.0 ± 0.2	– 0.3	[14]
apo-oTF/2N	22.0 ± 0.2		[14]
Fe^{3+}-oTF/2N	20.1 ± 0.2	– 1.9	[14]
apo-oTF/2C	21.8 ± 0.2		[14]
Fe^{3+}-oTF/2C	20.8 ± 0.2	– 1.0	[14]

[a] R_g = radius of gyration; $\Delta R_g = R_g - R_g(\text{apo})$.

have shown that binding of Fe^{3+} and carbonate causes the N-lobe to change its conformation from wide-open to closed [4–9, 7, 12]. This involves a 54° rotation of the N2 domain relative to the N1 domain and flexing of the two antiparallel extended polypeptide strands which run behind the iron binding site connecting domains N1 and N2. Figure 3 shows the open and closed forms of the N-lobe of human lactoferrin. The most remarkable change is to helix 5 in domain N2 which appears to pivot about helix 11. Small-angle X-ray scattering is a good technique for characterising the overall size and conformation of biological macromolecules in solution, and it has been shown that loading of transferrin with Fe^{3+}, Cu^{2+}, or In^{3+} causes a decrease in the radius of gyration consistent with lobe closure (Table 2) [13–15]. The isolated N and C-lobes undergo structural changes in solution on iron binding and release, similar to those of the corresponding lobes of intact transferrin, despite the fact that the apo C-lobe can open only to the equivalent of about 75% of the N-lobe due to the presence of an extra disulphide bridge [14].

The mechanism for opening and closing the lobes of transferrin may involve a pH-sensitive interdomain interaction. Endocytosis into acidic endosomes (pH ca. 5.5) could result in the protonation of one or both of residues Lys209 and Lys301, the so-called dilysine "trigger" (Fig. 4), based on the X-ray structure of the N-lobe of hen ovotransferrin [16]. The two resulting positive charges on opposite domains may then provide the driving force to push the two domains apart and hence allow release of the metal ion. A similar trigger is also possible for the N-lobe of human transferrin [16]. The crystal structure of diferric hen ovotransferrin shows that a related interdomain interaction, between Gln541 and Lys638, is also found in the C-lobe [8]. The distance between the oxygen atom of Gln541 and the nitrogen atom of Lys638 is ca. 3.0 Å. Both the Lys209-Lys301 and Gln541-Lys638 couples are near the iron binding sites and share the same position in each lobe, and both of them are also in

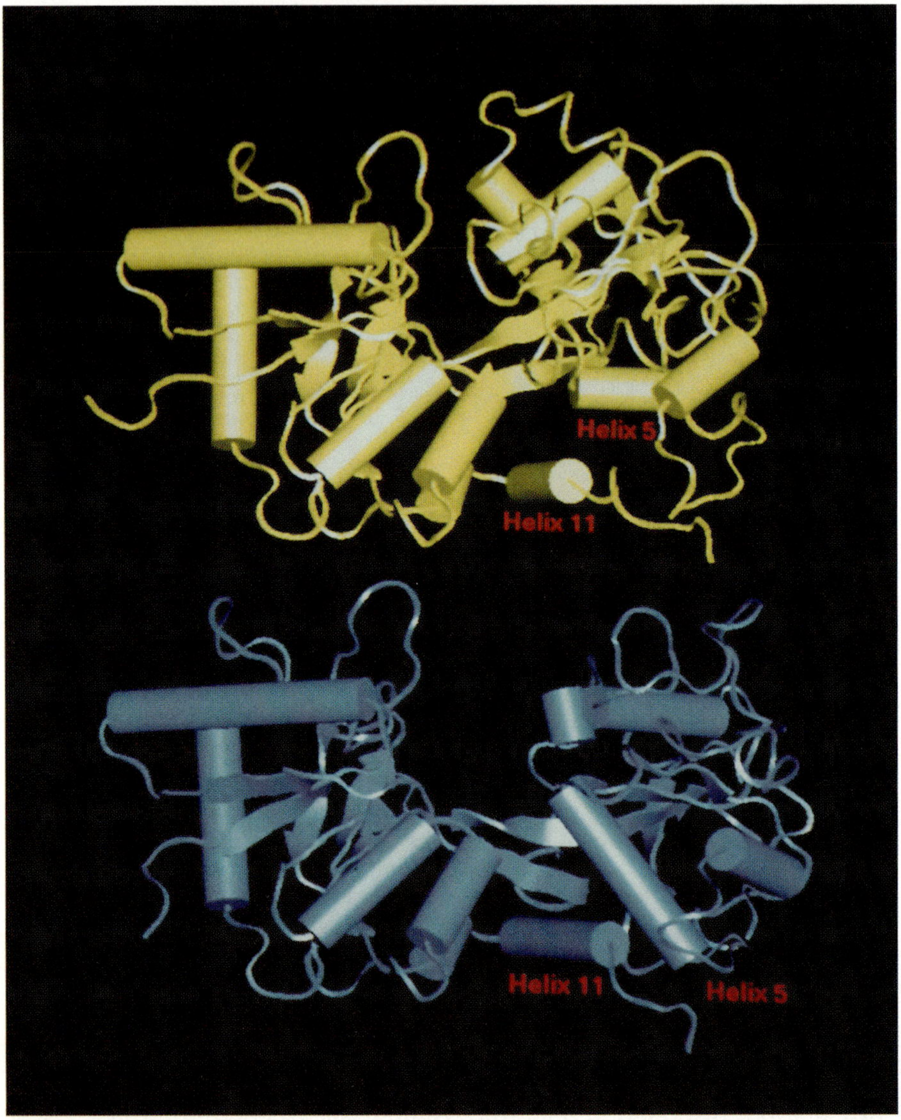

Fig. 3. Open (blue) and closed (yellow) forms of the N-lobe of human lactoferrin. Note the movement of helix 5 (domain N2) which appears to pivot about helix 11. The models were constructed using the program Quanta and are based on coordinates from Refs. [4, 7]

hydrophobic environments. That the C-lobe of ovotransferrin retains Fe^{3+} at lower pH values than the N-lobe is probably relevant to the difference in the two "triggers". Low-angle X-ray scattering studies of the N-lobe recombinant human serum transferrin have suggested that Asp63, an iron binding ligand, serves as a trigger for the closure of the two domains upon Fe^{3+} uptake, and that

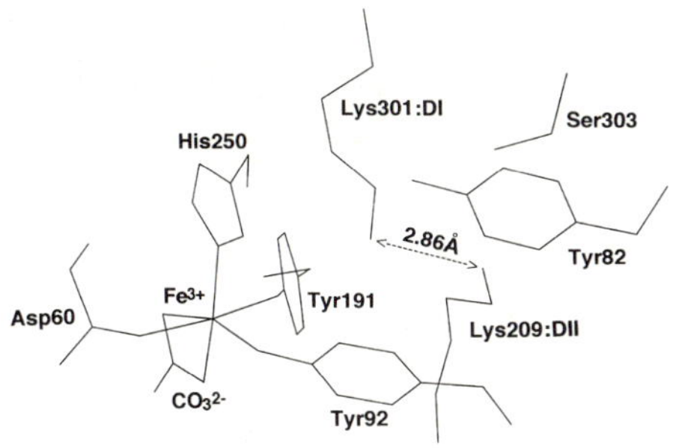

Fig. 4. The dilysine trigger. Part of Fe-N-lobe ovotransferrin showing the proposed [16] H-bonding between the ε-amino groups of Lys209 and Lys301 (one protonated and one deprotonated). Protonation of this pair may act as a trigger for lobe opening and metal release. For corresponding amino acid residue numbers for hTF see Fig. 2

this trigger is abolished completely by the mutation of Asp63 to Ser or Cys (i.e. the lobe does close) [17]. However recent studies of D63S lactoferrin by X-ray crystallography have shown that the N-lobe is completely closed which has led to the proposal of an equilibrium between open and closed forms in solution with a low energy barrier [18]. In addition the binding of another anion, such as chloride, is a prerequisite for Fe^{3+} release, even in the presence of a receptor [19].

3
Detection of Metal Binding

The transferrins are primarily iron-binding proteins, but in human serum, transferrin is only about 30% saturated with iron, so there is potential capacity for binding to other metal ions which enter the body. Indeed about 30 metal ions have been reported to bind to transferrin with either carbonate, oxalate or other carboxylates as synergistic anions, although Fe^{3+} has a higher affinity than any other metal ion for which the binding constant has been determined [20]. Other ions which bind include main group metal ions, such as Bi^{3+}, Ga^{3+}, In^{3+}, Al^{3+} and Tl^{3+}, transition metal ions such as Mn^{2+} [21], Cu^{2+}, Ni^{2+} and Ru^{3+}, and lanthanide metal ions such as La^{3+}, Ce^{3+}, Nd^{3+}, Sm^{3+} and Gd^{3+} [20–25]. Such binding may play an important role in the transport and delivery of medical diagnostic radioisotopes such as $^{67}Ga^{3+}$ and $^{111}In^{3+}$ [26], toxic metal ions such as Al^{3+} [27], and therapeutic metal ions such as Ru^{3+} [28, 29]. Transferrin is also thought to transport toxic transuranium elements such as Pu^{4+} in the body [30]. In rabbit plasma, Sc^{3+} has been found to be present as a transferrin complex both in vivo and in vitro [31]. Many metal ions have also been exploited for their spectroscopic properties in investigations of transferrin structure and function, although even if the metal binds to the iron site, the structure of the site may be perturbed to suit the metal ions.

A useful technique for the study of interactions between metal ions, anions and apotransferrin is electronic absorption spectroscopy. Apotransferrin is a colourless protein with an intense ultraviolet absorption near 280 nm with an ε_{278} of 93,000 M^{-1} cm^{-1}. This band is due to π-π^* transitions of the aromatic amino acids tyrosine, tryptophan and phenylalanine. When the phenol group of tyrosine is deprotonated (as it is when bound to a metal ion), two absorptions near 295 and 240 nm appear, allowing difference spectra to be recorded. This method not only distinguishes whether binding occurs at the specific iron sites, or at other sites (e.g. Pt^{2+}) [32], but also provides a method for determination of the strength of metal binding to transferrin (Sect. 5.1). Different anions can be distinguished due to slightly different effects on the absorption spectrum [33]. However, UV difference spectra do not give any distinct evidence for closed (after metal binding) or open (apo-form) lobe conformations, or information about whether Asp and His ligands are coordinated to bound metal ions.

For some metal ions, intense tyrosinate-to-metal charge-transfer bands in the visible region of the spectrum (400–500 nm; ε ca. $4-9\times10^3$ M^{-1} cm^{-1}) are diagnostic of site-specific binding to transferrin. For example the Fe^{3+} complex is orange-red with a band at ca. 465 nm, the Cu^{2+} and Co^{3+} complexes are yellow, and the Mn^{3+} complex brown [34].

Electron paramagnetic resonance (EPR) spectroscopy has been used to study Fe^{3+}, Cu^{2+}, Cr^{3+} and VO^{2+} transferrin complexes. In the case of Cu^{2+} and VO^{2+} it is possible to distinguish between the N- and C-lobe sites on the basis of the g values and hyperfine coupling constants, but for the other ions spectra can be difficult to interpret fully [35].

Interactions between both imidazole (His) and a ^{13}C-labeled synergistic anion with Cu^{2+} in ovotransferrin have been detected by electron spin-echo spectroscopy [36].

Recently NMR has been widely used for probing metal-transferrin complexes. Different NMR techniques can be used to investigate anions directly bound to the metal (^{13}C NMR) or metal in the specific iron sites (multinuclear NMR), and metal-induced structural changes to the protein. The order of lobe-loading can also be determined (Sect. 4).

Extended-X-ray-absorption-fine-structure (EXAFS) spectroscopy is a technique that can be used to study differences in the metal-binding sites and also to provide accurate information on the metal environment [37–40]. Average metal-ligand bond lengths and coordination numbers can be determined, but O cannot be distinguished from N as a ligand and the two metal sites in M_2-transferrin complexes cannot be distinguished.

X-ray crystallography is currently the most powerful method for the study of the structure of transferrin. The structures of several transferrins with different bound metal ions and different anions have been solved [3–9, 11, 12, 41–45] (Table 1). The polypeptide folding and domain closure for Fe^{3+} lactoferrin are similar with either carbonate or oxalate as synergistic anion. Each lobe contains an approximately octahedral Fe^{3+} binding site. Oxalate binds to Fe^{3+} in symmetrical 1,2-bidentate fashion (Fe-O distances 2.07 and 1.91 Å) in the C-lobe whereas in the N-lobe the anion coordination is remarkably asymmetrical (2.55 and 1.87 Å). The crystal structure of Cu^{2+}-substituted lactoferrin shows that although the overall structure of the protein is unchanged compared with diferric lactoferrin, the metal sites are subtly different. In the N-lobe, Cu^{2+}

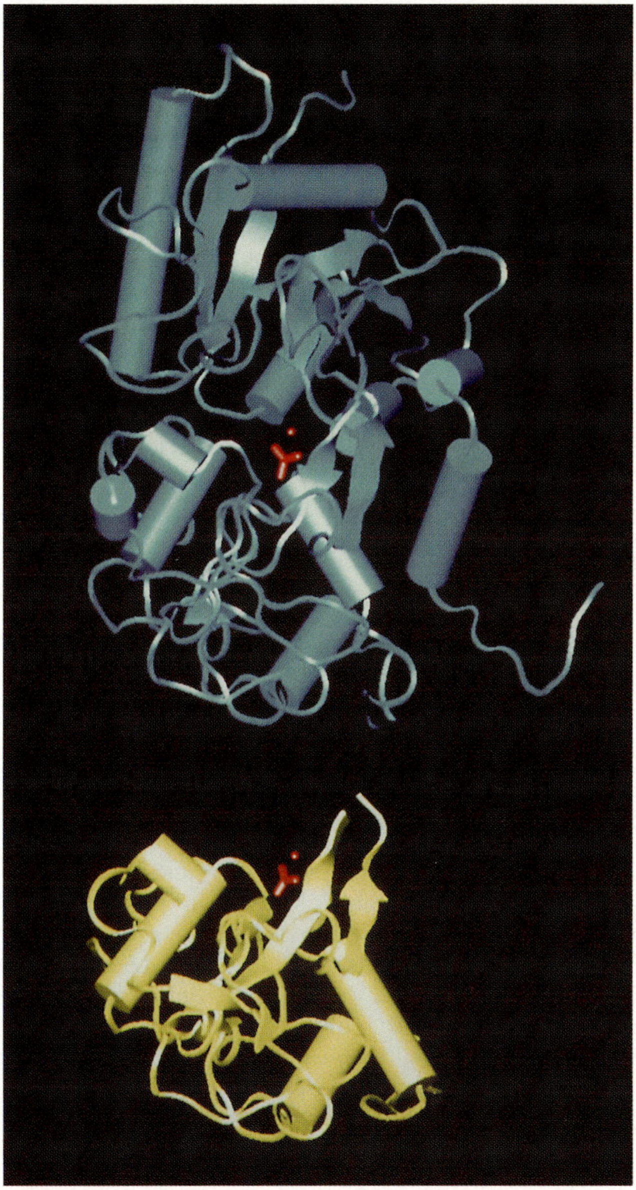

Fig. 5. Comparison of Fe^{3+} binding sites in the N-lobe of ovotransferrin (blue) and an 18 kDa fragment of domain N2 of duck ovotransferrin (yellow) [based on Ref. 45]. Iron and carbonate are shown in red. In the N2-domain structure, Fe^{3+} is bound to the protein via the two tyrosine residues only

ion is pentacoordinate and approximately square-pyramidal, with a long apical bond of 2.8 Å to Tyr92 and a monodentate carbonate ion. In the C-lobe, however, Cu^{2+} is hexacoordinate but more distorted from regular octahedral geometry. Carbonate can be displaced from the $Cu_2(CO_3)_2LF$ ternary complex by addition of a large excess of oxalate, but no such displacement occurs for $Fe_2(CO_3)_2LF$ [43]. In Cu_2(oxalate)LF, Cu^{2+} is also pentacoordinate in the N-lobe bound to a monodentate oxalate (2.0 Å).

The structures of complexes with Fe^{3+} bound to the isolated N-lobe of transferrin (half molecule) and N2 domain (quarter-molecule) have also been solved by X-ray crystallography [44–46]. The former shows that both the protein structure and the metal and anion binding site are the same as in the intact protein. The latter shows that the anion binding site, which is formed by residues of the N2 domain, is unchanged, and that Fe^{3+} is bound to two Tyr residues (Tyr95 and Tyr188) and a bidentate carbonate and two other unknown ligands (possibly glycine). The folding of this fragment is the same as for the N2 domain of intact transferrin (Fig. 5). This has allowed a proposal to be made [45] for the mechanism of Fe^{3+} uptake by transferrin: binding of synergistic anion (carbonate), followed by binding of the metal to the lobe in the open form, and then closure of the domains with formation of interdomain hydrogen-bonding (Asp trigger).

4 Determination of the Order of Lobe-Loading

A useful technique for determining the order of lobe-loading is electrophoresis in 6 M urea as described by Makey and Seal [47–49] (Table 5). This involves the separation of M_2-TF from M_C-TF and M_{N}-TF and works well providing that the metal ion remains bound to the protein under the conditions of electrophoresis, which is not always the case (e.g. for Al^{3+}). Another method is differential scanning calorimetry (DSC). Apo-hTF shows two major thermal transitions with T_m values of 57.6 and 68.4 °C, respectively, whereas ovotransferrin shows only a single transition at 60 °C. Addition of one mol equiv of Fe^{3+} (as $[Fe(NTA)_2]$) to hTF solution perturbs the transition temperature of only one lobe. Addition of the second mol equiv Fe^{3+} affects the transition temperature of the other lobe. By comparing the results with those of the isolated N-lobe, it was determined that the C-lobe was preferentially loaded. Although promising, there are only two studies on other metals binding to transferrin using this method [50, 51]. EPR has also been used to determine the order of lobe-loading of paramagnetic ions such as VO^{2+} [52] and Gd^{3+} [25, 53, 54].

Recently, various NMR methods have been used to detect the order of lobe-loading of metal ions. ^1H NMR studies of transferrin are complicated by the severe overlap of signals and the broadening of the resonances (slow tumbling due to large M_r). But with the aid of resolution enhancement, resonances for most of the His residues can be resolved and pH titration curves have been established [55]. In the high field region of the spectrum (ca. 0.5 to −2 ppm), peaks due to aromatic ring-current shifted methyl groups can be resolved. The sharpest peaks in the spectrum (2.0 ~ 2.3 ppm) belong to the N-acetyls of

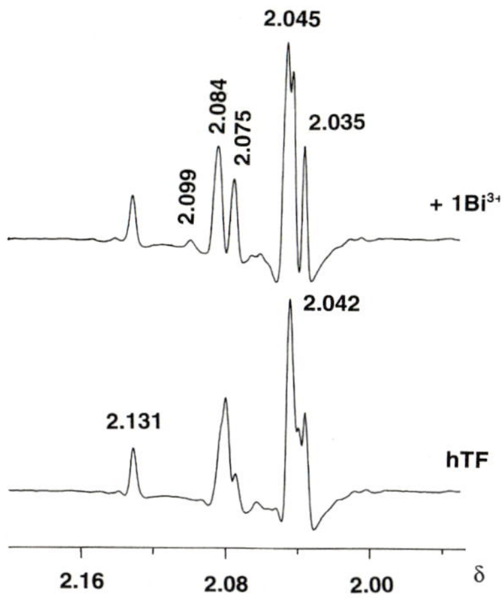

Fig. 6. Use of glycan chain ^{1}H NMR resonances for monitoring metal binding to the C-lobe. 750 MHz ^{1}H NMR spectra showing the N-acetyl peaks of the C-lobe glycan chains of human serum transferrin before and after addition of 1 mol equiv of Bi^{3+} with bicarbonate as the synergistic anion. No further changes occur on binding the second equivalent of Bi^{3+} suggesting that Bi^{3+} binds preferentially to the C-lobe

the glycan chains, attached to Asn416 and Asn611 in the C-lobe. The role of the carbohydrate moieties is still unclear. They appear to play no role in transferrin binding to cell receptors [56]. Metal ions bound to the C-lobe can cause chemical shift changes in this region probably due to structural perturbations (Fig. 6). Al^{3+}, Ga^{3+}, In^{3+} and Bi^{3+} all bind tightly to apotransferrin and resonances for free and bound forms are usually in a slow exchange on the NMR time-scale when these metal ions are titrated into solutions of the protein. By comparing the pattern of shift changes for transferrin itself with those of the isolated recombinant N-lobe, the order of lobe-loading can be determined [57–59].

4.1 NMR of Synergistic Anions

Since the synergistic anion is bound directly to the metal, it is possible to detect the lobe-loading via ^{13}C-NMR studies of ^{13}C-labeled anions (e.g. $H^{13}CO_3^-$ and $^{13}C_2O_4^{2-}$). The ^{13}C chemical shift for bound anion in each lobe is slightly different, and by comparing the chemical shift with that of the recombinant N-lobe transferrin, the order of lobe-loading can be determined (Table 5). Sometimes spin-spin coupling between the metal and ^{13}C (from the synergistic anion) can be observed, such as ^{2}J (^{13}C-^{205}Tl) 270–290 Hz ($^{13}CO_3^{2-}$) and 15–30 Hz ($^{13}C_2O_4^{2-}$). Non-equivalence of the ^{13}C atoms in oxalate results in a ^{1}J (^{13}C-^{13}C) value of ca 70–75 Hz ($^{-}O_2{}^{13}C$-$^{13}CO_2^-$), and provides further evidence for oxalate being directly bound to the metal ion [60, 61].

4.2 Metal NMR

Multinuclear NMR spectroscopy using diamagnetic metal ions can provide a direct method for determining the order of lobe-loading. Thus for ^{205}Tl ($I = 1/2$), there is a 21 ppm difference between resonances for Tl^{3+} in the N- and C-lobes of ovotransferrin, although for the quadrupolar nucleus ^{27}Al there is no shift difference for serum transferrin but a 1.5 ppm difference (11.7 T) for ovotransferrin. Other nuclei such as ^{45}Sc, ^{71}Ga, ^{51}V and ^{113}Cd have also been used to monitor lobe-loading [62–65], and these data are summarised in Table 3.

Resonances of half-integer quadrupolar nuclei bound to transferrin are much sharper when recorded at very high magnetic fields. In the limit of slow isotopic molecular motion, the linewidth ($\Delta\nu_{1/2}$) of the central transition ($m = 1/2$ to $-1/2$) decreases with increasing magnetic field:

$$\Delta\nu_{1/2} = k[\chi^2/(\tau_c \nu_o^2)] \tag{1}$$

where χ is the quadrupolar coupling constant ($\chi = e^2qQ/h$), τ_c is the correlation time for fluctuations in the electric field gradient at the nucleus and ν_o is the resonance frequency which is proportional to the magnetic field. It has also been shown that decreasing temperature and increasing viscosity give rise to a decrease in line widths [66]. Caution must be exercised in these quadrupolar systems, the chemical shift is field-dependent, and the second-order dynamic frequency shift is given by [67, 68]:

$$\Delta\delta_d = k(\chi^2/\nu_o^2)\,. \tag{2}$$

Table 3. Chemical shift differences ($\Delta\delta$ ppm) for various metal nuclei bound to the N- and C-lobes of serum transferrin or ovotransferrin.

Nucleus	Transferrin	Synergistic anion	$\Delta\delta$ (magnetic field)[a]	Ref
$^{27}Al(III)$	oTF	carbonate	2.2 (9.4 T); 1.5 (11.7 T)	[66]
		oxalate	2.0 (9.4 T); 1.3 (11.7 T)	
	hTF	carbonate	0	
$^{45}Sc(III)$	oTF	carbonate	8.0 (11.7 T)	[62]
	hTF	carbonate	0 (11.7 T)	
$^{51}V(V)$	oTF	[b]	0	
	hTF	[b]	– 1.5 (9.4 T); – 1.5 (11.7 T)	[64]
$^{71}Ga(III)$	oTF	carbonate	25 (11.7 T)	[63]
		oxalate	31 (11.7 T)	
$^{113}Cd(II)$	hTF	carbonate	– 2.0	[65]
$^{205}Tl(III)$	oTF	carbonate	21	
		oxalate	3	
	hTF	carbonate	17	[60, 103]

[a] $\Delta\delta = \delta$(N-lobe) – δ(C-lobe); the chemical shifts of quadrupolar nuclei are field-dependent.
[b] VO_2^+ binds without a requirement for a synergistic anion.

For ^{71}Ga (I = 3/2) in ovotransferrin the chemical shift is -103 ppm at 11.7 T but −57 ppm at 17.6 T [63]. These multinuclear NMR experiments on quadrupolar nuclei can require long spectral accumulation times.

4.3 Isotopic Labelling of Protein

Isotopic labelling is a promising method for providing the assignments of resonances for specific amino acid residues of transferrin. By means of 2D [^{1}H, ^{13}C] HMQC-NOESY and molecular modelling, the S-$^{13}CH_3$ resonances of all the five Met residues of the N-lobe and nine Met residues of intact transferrin have been tentatively assigned [69, 70]. The advantage of choosing to label the ε-CH_3 group of Met is that each ^{1}H resonance is a singlet and also Met has a low abundance in proteins. Some of the Met resonances (e.g. from Met464 and Met109) are sensitive to metal binding even though these residues are far away from the metal binding site, probably due to structural changes in the protein induced by metal ions. The order of lobe-loading can be easily detected via 2D [^{1}H, ^{13}C] heteronuclear multiple quantum coherence (HMQC) spectra acquired within 1 h at transferrin concentrations of only ca. 0.3 mM. The 2D HMQC method involves inverse detection of ^{13}C, i.e. detection of ^{13}C via observation of

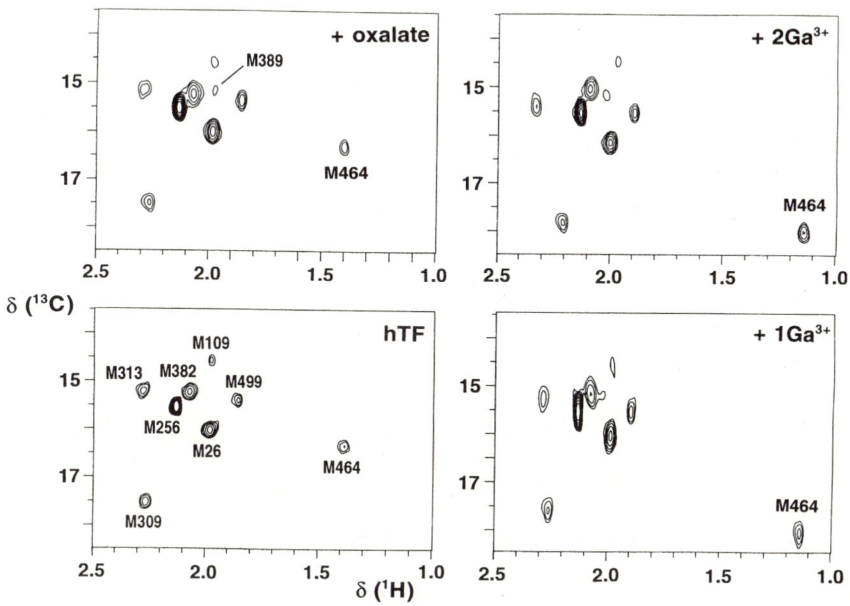

Fig. 7. Use of ^{13}C-labeled transferrin to determine the order of lobe loading. 2D [^{1}H, ^{13}C] HMQC spectra of apo-[ε-^{13}C]Met-hTF before, and after addition of 10 mol equiv of oxalate, followed by addition of 1 mol equiv and 2 mol equiv of Ga^{3+}. Specific shifts of cross-peaks on binding either the first or second equivalent of Ga^{3+} indicate the order of lobe-loading; e.g. Met464 in the C-lobe shifts on binding the first equivalent but not on binding the second (data from reference [69])

^1H resonances and is much more sensitive (by up to $(\gamma_H/\gamma_{13C})^{5/2}$) than direct ^{13}C detection. Figure 7 shows the HMQC spectrum of hTF before and after addition of Ga^{3+}. This method has also been used to determine the order of lobe-loading of Fe^{3+} and other trivalent metal ions [71] (Table 5). Due to the ^{13}C labelling of Met and the sensitivity enhancement by inverse detection, it should be possible to detect lobe-loading at concentrations of biological relevance (ca. 35 μM).

5 Transferrin-Metal Binding Constants

5.1 Determination of Binding Constants

The complexation of metal ions to the phenolic groups of the tyrosine residues in the specific metal-binding sites of apo-transferrin leads to the production of two new absorption bands at ~ 240 nm and ~ 295 nm in UV-difference spectra. This has been widely used to determine the number of metal ions that are bound and to determine metal binding constants.

Typical UV difference spectra of metal-transferrin and titration curves are shown in Fig. 8. For some metal ions which are readily hydrolysed at biological pH, such as Bi^{3+}, Ga^{3+}, In^{3+} and Al^{3+}, chelating ligands such as NTA and EDDA have been chosen as complexing ligands to maintain the metal ions in solution

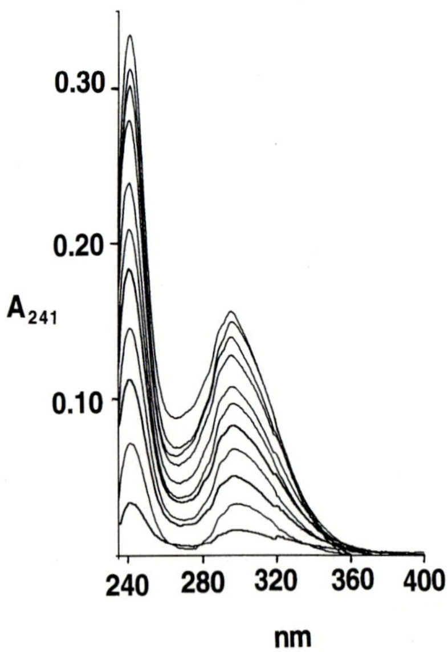

Fig. 8. Detection of metal binding to transferrin by UV difference spectroscopy. The bands arise from the deprotonation of phenol groups of Tyr residues of hTF on metal binding (in this case addition of increasing amounts of Bi^{3+}). The spectra are shown as a superimposed series, corresponding to successive increments of the added metal ion (Bi^{3+}).

and produce a concentration of free metal ion sufficient to partially load the two metal-binding sites of the protein.

The method frequently used by Harris and co-workers [21, 23, 27] for calculation of metal-transferrin binding constants is to measure the absorbance at ~ 240 nm and fit the experimental data by an interactive procedure to minimise the difference between calculated and analytical values of $\Delta\varepsilon$,

$$\Delta\varepsilon_{\text{calcd}} = \Delta\varepsilon_M K_1[M][TF] + 2\,\Delta\varepsilon_M K_1 K_2 [M]^2[TF] \tag{3}$$

where $\Delta\varepsilon_M$ is the molar absorptivity per bound metal ion for metal transferrin, obtained from the initial linear portion of the curve. For a given set of experiments, the concentrations of free metal ions, free competing ligands and apotransferrin are evaluated by iteration to minimise the difference between calculated and experimentally fixed values of total metal, total ligand and total transferrin. By assuming a set of values for K_1, K_2 and K_{ML}, the values of K_1, K_2 and $\Delta\varepsilon_M$ are then varied to minimise the sum of the squares of the residuals between the observed and calculated $\Delta\varepsilon$ values. Several metal-transferrin binding constants have been calculated in this way.

We have found that for metal ions with known stability constants for chelating ligands, a method utilising a Hill plot is effective for the calculation of metal-transferrin binding constants, especially for K_1 [22]. The method is based on the calculation of equilibrium constants for a competition reaction between transferrin and chelating ligands (K_{a1} and K_{a2}), such as NTA and EDDA. Metal-transferrin stability constants are then obtained by simply multiplying the stability constant of the metal chelate complex (K_{ML}) by the stability constant of the metal complex and apotransferrin. By assuming that the two binding sites of transferrin are independent and equivalent, the following equation can then be used to fit the experimental data:

$$1/Y = 1/n + 1/(nK_a)\,[L]/[ML] \tag{4}$$

where K_a is the intrinsic binding constant, $K_{a1} = 2K_a$ and $K_{a2} = K_a/2$, and $Y = [\text{M-bound}]/[\text{hTF}]_{\text{total}}$. The concentration of metal bound to transferrin is calculated by assuming that fully loaded metal-transferrin has a molar absorptivity of $\Delta\varepsilon_1$. By choosing different ratios of chelating ligands (L) to metal ions, a single metal site of transferrin can be loaded and the equation above simplifies to:

$$1/Y = 1 + 1/K_{a1}[L]/[ML] \tag{5}$$

This method has been used to calculate Bi-hTF and Sc-hTF binding constants successfully [22, 72]. The advantage of this method is that the accuracy of the stability constants can be estimated not only from different titration curves but also from the intercepts of the plots. The second binding constant is then recalculated using the literature method [73] giving rise to two binding constants for the binding sites of transferrin. After obtaining K_1, the method is extremely sensitive to K_2 as shown in Fig. 9.

Other methods, such as equilibrium dialysis and EPR, have also been used to calculate the stability constants of metal-transferrin complexes (e.g. Fe^{3+}, Gd^{3+} and Mn^{2+}), more details can be found in a previous review [35].

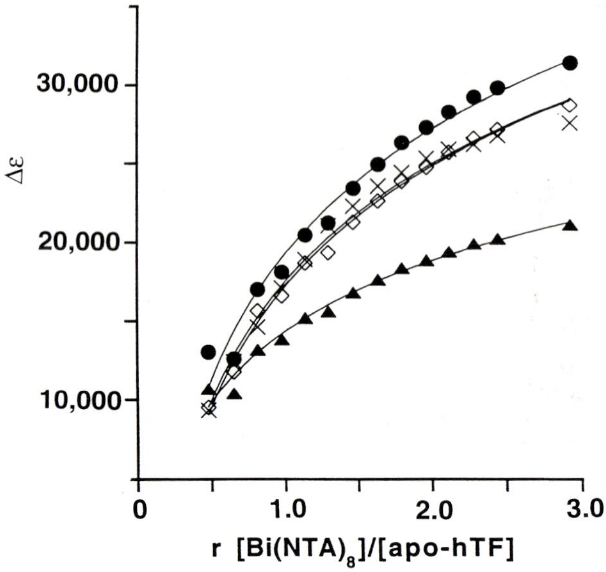

Fig. 9. Example of the determination of the second binding constant of human transferrin for metal ions (in this case Bi^{3+}) by simulation of the change in the extinction coefficient (at 241 nm) with metal to apotransferrin ratio. Key: x, experimental points; ●, log K_2 19.00; ▲, log K_2 18.00; ◊, log K_2 18.80 (best fit). Data from Ref. [22].

5.2 Differences Between Lobes

Transferrin contains two metal-binding sites with remarkably similar thermodynamic and spectroscopic properties [74–76]. Fletcher and Huehns [77, 78] and Harris and Aisen [79] have shown that the two sites are different in their ability to donate iron to reticulocytes. The equilibrium studies of Aasa et al. [74] and Aisen and Leibman [80] demonstrated the presence of two sites a and b on transferrin, which were identified as the C- and N-terminal sites, respectively [81], and which are different kinetically and thermodynamically. Aisen et al. [82] have shown that the C-terminal site of transferrin binds Fe^{3+} more strongly than the N-terminal site, by a factor of about 20 at pH 7.4, with an ambient bicarbonate concentration. Similar behaviour has also been found for other metal ions, such as Ga^{3+} and Bi^{3+}: the two lobes differ by a factor of 10 [73] and 6.8 [22], respectively, and for In^{3+}, the factor is 76 [83]. Normally the two sites differ by approximately one unit in their log K values, which is beyond the purely statistical factor of 4, primarily due to a difference in the intrinsic binding affinities of the two lobes. In vitro, the two lobes of transferrin display somewhat different Fe^{3+} release properties. The N-lobe releases Fe^{3+} at a pH of about 5.7, while the C-lobe retains Fe^{3+} down to pH values of about 4.8 [84–86]. Binding of Fe, Co-transferrin to the transferrin receptor at pH 5.6 apparently reverses the order of metal stabilisation: Fe^{3+} is labilised (by pyrophosphate) from the C-lobe site in preference to the N-lobe [87].

5.3 Influence of Anion and Conditions

When transferrin binds metal ions, it binds an anion, the so called "synergistic anion", cooperatively. This is bicarbonate in the normal biological system. Some carboxylates such as oxalate, malate and lactate can also serve this function. Thus the stability constants of metal-transferrins are affected by the concentration of bicarbonate. The binding constant of apo hTF for Ga^{3+} (log K_1) is 18.1 at an ambient bicarbonate concentration (0.14 mM) and 20.3 at the serum bicarbonate level (27 mM) [73]; and similarly for Fe^{3+} log K_1 is 20.7 and 22.8, respectively [80]. Binding therefore becomes stronger at higher bicarbonate concentrations. In order to compare the metal-hTF binding constants which have been measured at a different bicarbonate concentration, the following relationships have been verified [88]:

$$\log K^* = \log K_M + \log \alpha_c \tag{6}$$

where K^* is the bicarbonate-independent stability constant, $\alpha_c = K_c[HCO_3^-]/(1 + K_c[HCO_3^-])$, and K_C and K_M represent equilibrium constants of the following two reactions

$$K_C\text{: } HCO_3^- + \text{apoTf} = HCO_3\text{-TF} \tag{7}$$

$$K_M\text{: } M^{3+} + HCO_3^-TF = M\text{-(H)}CO_3\text{-TF.} \tag{8}$$

Non-synergistic anions such as Cl^-, ClO_4^-, $HP_2O_7^{3-}$ and ATP^{3-} have also been reported to affect the thermodynamic stability of metal-transferrin complexes. Chasteen and co-workers [89–91] proposed non-synergistic anion binding based on the EPR spectral measurements. Williams et al. [92] reported that the thermodynamic stability of iron binding to the *N*-terminal site relative to the C-terminal site increased with increasing concentrations (0 to 0.5 M) of NaF, NaCl, NaBr, NaI, $NaNO_3$, Na_2SO_4 and $NaClO_4$, and that this effect was enhanced as the pH increased in the range 7 to 8.4. The nature of the interaction of the non-synergistic anions with transferrin is unclear, but it does affect the structure and reactivity of the metal-binding sites [35].

For different kinds of metal ions with different anions as synergistic anions, pH-dependent binding behaviour can be different. In this aspect Cu^{2+}-transferrin has been studied in much more detail. It has been found [93] that Cu^{2+} does not bind to serum transferrin below pH 6 , and that Cu^{2+} binds only to the C-terminal site as the pH is raised from 6.0 to 7.2. At higher pH it binds also to the N-terminal site. On further increasing the pH to above 9.5, Cu^{2+} is lost from the specific metal-binding sites [94].

6 Rationalisation of Transferrin Metal-Binding Constants

It has been suggested that the strength of binding of trivalent metal ions to transferrin is related to the size of the metal ion, being optimum for Fe^{3+} (ionic radius 0.65 Å), weaker for slightly smaller (Ga^{3+}, 0.62 Å) or larger (In^{3+}, 0.80 Å)

Table 4. Stability constants for metal binding to serum transferrin (log K)[a], metal radii, and molar absorptivity per bound metal ion ($\Delta\varepsilon_1$)

Metal ion	radius (Å)	$\Delta\varepsilon_1$ ($M^{-1}cm^{-1}$)	[HCO_3^-]	log K_1	log K_2	Refs
Fe^{3+}	0.65	18,000	Ambient	21.44	20.44	[82]
Bi^{3+}	1.03	21,900	5 mM	19.42	18.58	[22]
Ga^{3+}	0.62	20,000	5 mM	19.77	18.82	[98]
In^{3+}	0.80	17,200	5 mM	18.52	16.64	[83]
Sc^{3+}	0.75	22,000	5 mM	14.60	13.30	[72]
Al^{3+}	0.54	14,800	5 mM	13.5	12.5	[27]
Gd^{3+}	0.94		Ambient	6.8		[25]
Sm^{3+}	0.96	21,000	Ambient	8.35	6.61	[23]
Nd^{3+}	0.98	18,700	Ambient	7.31	5.26	[23]
Ni^{2+}	0.69	14,800	5 mM	4.10	3.23	[21]
Zn^{2+}	0.74	13,300	15 mM	7.8	6.4	[88]
Fe^{2+}	0.78			3.2[a]	2.5	[21]
Mn^{2+}	0.83	10,100		4.06		[73]
Cd^{2+}	0.95	11,600		6.19	5.10	[99]

[a] Estimated from a linear-free-energy relationship (LFER) for the complexation of Ni^{2+} and Fe^{2+}.

ions, and much weaker for very small (Al^{3+}, 0.54 Å), or very large ions (lanthanides, 0.86 – 1.03 Å)(see Fig. 10 and Table 4) [95]. It has been assumed for this "Venus Fly-Trap" protein that for strong binding the size of the metal ion must be matched to the size of the interdomain binding cleft, and this argument has been used to rationalise the recent finding that lactoferrin promotes oxidation of the large metal ion Ce^{3+} (1.10 Å) to the smaller Ce^{4+} (0.87 Å) [96].

However, recently it has been found that the large metal ion Bi^{3+} (ionic radius 1.03 Å), an ion widely used in medicine as an anti-ulcer drug [97], binds very tightly to human transferrin, whereas if the strength of binding of metal ions to transferrin is optimised to the size of the interdomain binding site, then Bi^{3+} should bind as weakly as the lanthanide ions. In fact Bi^{3+} binds almost as strongly as Ga^{3+}, with a log K_1 of 19.42 and a log K_2 of 18.58 [22]. These findings have given rise to a new proposal for rationalisation of binding constants for both trivalent and divalent metal ions to transferrin and for an extension to other proteins, such as carbonic anhydrase and carboxypeptidase. Linear correlations between the binding constants of metal ions for proteins and small ligands allow discussion also of the preorganisation of metal sites in proteins.

6.1 Correlations Between the Strength of Metal Binding to hTF and Small Ligands

It has been noted that linear-free-energy-relationships (LFER) exist between the strengths of binding of Fe^{3+} and other metal ions to transferrin and binding to a range of oxygen and nitrogen donor ligands [22, 27], and it seemed reasonable to propose [22] that the types of amino acid side-chain donors in

Table 5. Order of lobe loading for metal ions

Metal donor	Protein	Synergistic anion	pH	Method[a]	Preference	Reference
$(NH_4)_2Fe(SO_4)_2$	hTF	bicarbonate	7.4–8.5	E	N-lobe	[49]
Fe(NTA)	hTF	bicarbonate	6–8.5	E; Cal;	C-lobe	[48, 50, 51, 81, 82]
				NMR		[71]
Fe^{3+} citrate	hTF	bicarbonate	7.4–8.5	E	N-lobe	[49]
Fe^{3+} oxalate	hTF					[49]
Cr^{3+} citrate	hTF	bicarbonate	7.5	EPR	both	[54]
			5.9		N-lobe	[49]
$Al_2(SO_4)_3$	hTF	bicarbonate	8.8	NMR	N-lobe	[58]
$Al(NO_3)_3$	oTF	bicarbonate	7.6	NMR	N-lobe	[61]
	oTF	bioxolate	7.6	NMR	C-lobe	
Ga(NTA)	hTF	oxalate	7.25	NMR	C-lobe	[57, 69]
		bicarbonate	7.25	NMR	both	[71]
$GaCl_3$	oTF	bicarbonate	7.6	NMR	N-lobe	[63]
		oxalate	7.6	NMR	both	
In^{3+} citrate	hTF	bicarbonate	7.25	NMR	C-lobe	[59]
Sc(NTA)	hTF	bicarbonate	7.4	NMR	C-lobe	[72]
$ScCl_3$	oTF	bicarbonate	7.6	NMR	both	[62]
		or oxalate	7.4			
Bi(NTA)	hTF	bicarbonate	7.25	NMR	C-lobe	[22]
$CoCl_2 + H_2O_2$	hTF	bicarbonate	7.4	NMR	N-lobe	[53]
$GdCl_3$	hTF	bicarbonate	7.4–8.5	EPR	C-lobe	[25]
$VOSO_4$	hTF	bicarbonate	6.0	EPR	C-lobe	[52]
			7.5–9.0			
$TlCl_3$	hTF	bicarbonate	7.3–7.8	NMR	C-lobe	[103]
	oTF	bicarbonate	7.6		both	[60]

[a] E electrophoresis; Cal calorimetry.

transferrin determine the strengths of metal binding rather than the size of the binding cleft.

In Fig. 11 the constants ($\log K_1$) for binding to the first lobe of transferrin [83, 98, 99] are plotted against the stability constants ($\log K_1(OH^-)$) for hydroxide binding [100]. The second binding constant for transferrin is usually ca. 1 log unit lower than $\log K_1$, and $\log K_1(OH^-)$ values are related to pK_a values by $\log K_1(OH^-) = 14 - pK_a$. Hence the most readily hydrolysed (most acidic) metal ions bind most strongly to transferrin. Such a correlation between binding to transferrin and to RO^- ligands also holds for complexes of model phenolate ligands. Thus there is an excellent correlation with the binding constants for 1,2-dihydroxybenzene (Table 6). Therefore it is not surprising that Fe^{3+} binds to transferrin more strongly than most other metal ions. Other highly acidic metal ions such as Bi^{3+} and Tl^{3+} also bind strongly to transferrin and it is easy to

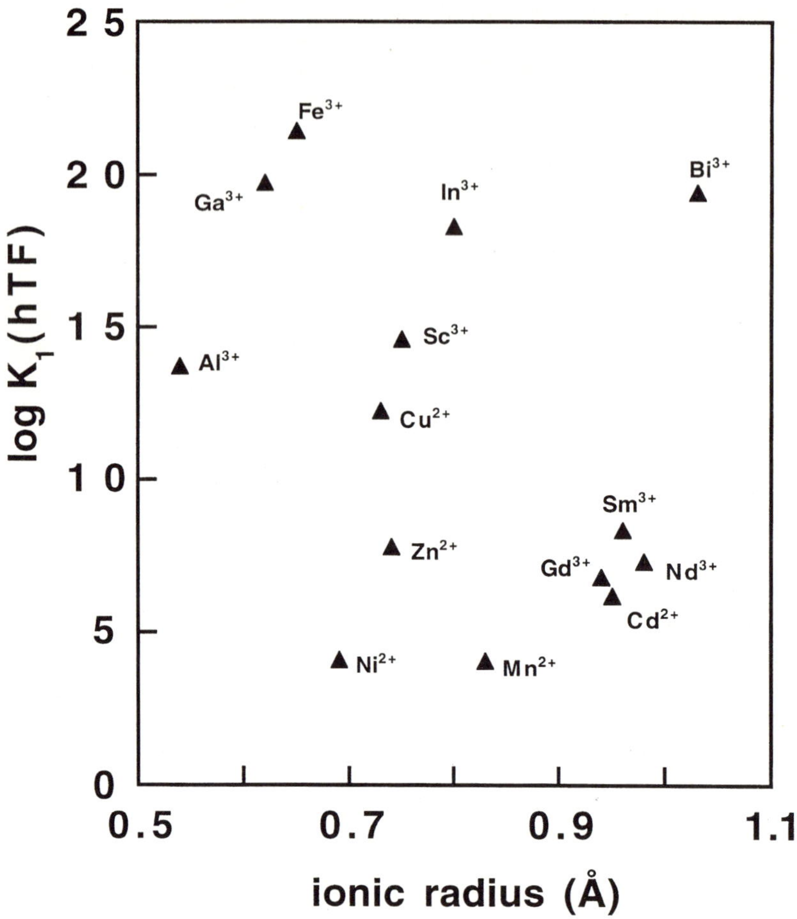

Fig. 10. Dependence of the first binding constant of human transferrin for divalent and trivalent metal ions on the ionic radius of the metal. The radii are those for coordination number six [95]. All stability constants were determined under similar conditions (10 ~ 100 mM Hepes buffer, 5 ~ 20 mM bicarbonate, 298 K, except Bi^{3+} at 310 K). Although some trivalent metal ions lie on a curve which appears to indicate that Fe^{3+} has an optimum size for strong binding, clearly many other ions do not follow this trend, and an alternative rationalisation is needed (see Fig. 11)

understand why trivalent metal ions bind to transferrin more strongly than divalent metal ions and why Cu^{2+} binds to transferrin more strongly than other divalent metal ions. Previous observations have suggested that most of the divalent transition metal ions bind only weakly to hTF. However Cu^{2+} is an exception, and Cu-hTF and Cu-Lf complexes have been used more than any other metal ion (except Fe^{3+}) for physicochemical studies of transferrins [41]. Recent data on Cu^{2+} transferrin (log K^*_1 12.25 for hTF) confirm that it binds less strongly than Al^{3+} and Mn^{3+} [101].

Table 6. Fitting parameters for linear correlations of the stability constants for the binding of metal ions to low molecular mass ligands (ligand (1)) and other low molecular mass ligands or proteins (ligand (2)). The values were obtained as described in the text and for low molecular ligands are based on stability constants in the Stability Constant Database [100]

Ligand (1)	Ligand (2)	(n)[a]	Intercept (Calc.)[b]	Slope	R[c]
OH	acetate	1	– 0.08 (0)	0.316	0.902
OH	oxalate	2	1.78 (1.74)	0.56	0.970
OH	citrate	3	3.44 (3.49)	0.84	0.924
OH	NTA	4	4.92 (5.23)	0.98	0.947
OH	EDTA	6	8.13 (8.72)	1.56	0.948
OH	hTF		– 3.37	2.05	0.977
catecholate	hTF		– 4.41	1.22	0.972
imidazole	CA		3.72	2.07	0.960
imidazole	CP		2.91	1.86	0.977

[a] n = denticity;
[b] calculated intercept = (n-1)log 55.5;
[c] Correlation coefficient.

We also examined other possible correlations between metal binding to transferrin and small N,O-donor ligands such as NTA, oxalate, acetate, glycine, malonate and lactate, but none of them were as good as that with RO^-. This suggests that the two tyrosinate ligands at the metal binding site of transferrin play dominant roles in determining the strength of metal binding.

6.2 Prediction of Transferrin-Metal Binding Constants

The correlation between the strength of metal binding to transferrin and metal acidity may provide a basis for prediction of other unknown stability constants for metal-transferrin complexes. This has been verified by the good agreement between the calculated (log K_1 14.6 ± 0.2) and predicted (log K_1 15.6 ± 1.6) stability constants for Sc^{3+}-transferrin [72]. Thus stability constants for other metal ions with transferrin can be predicted from Fig. 11 on the basis of their acidity. Tl^{3+} should bind to transferrin even more strongly than Fe^{3+}. Both ^{205}Tl NMR [102] and ^{13}C NMR [103] of the synergistic anions carbonate and oxalate have demonstrated that Tl^{3+} does indeed bind tightly, although the binding constants do not appear to have been determined.

A further prediction is that Cr^{3+} binds strongly to transferrin with log K_1 17 ± 1.6. The predicted binding constant for Co^{3+} (log K_1 21.4 ± 1.6) is similar to that for Fe^{3+}, and it is evident from reported work on Co^{3+}-transferrin complexes that binding is indeed strong even though the binding constant has not been reported [104]. La^{3+} and Yb^{3+} bind weakly to transferrin with log K_1 5.0 ± 1.3 and 8.7 ± 1.3, respectively, whereas Th^{4+} (ionic radius 0.94 Å) is known to bind very strongly to transferrin[105, 106], the predicted binding constant being log K_1 16.5 ± 1.5.

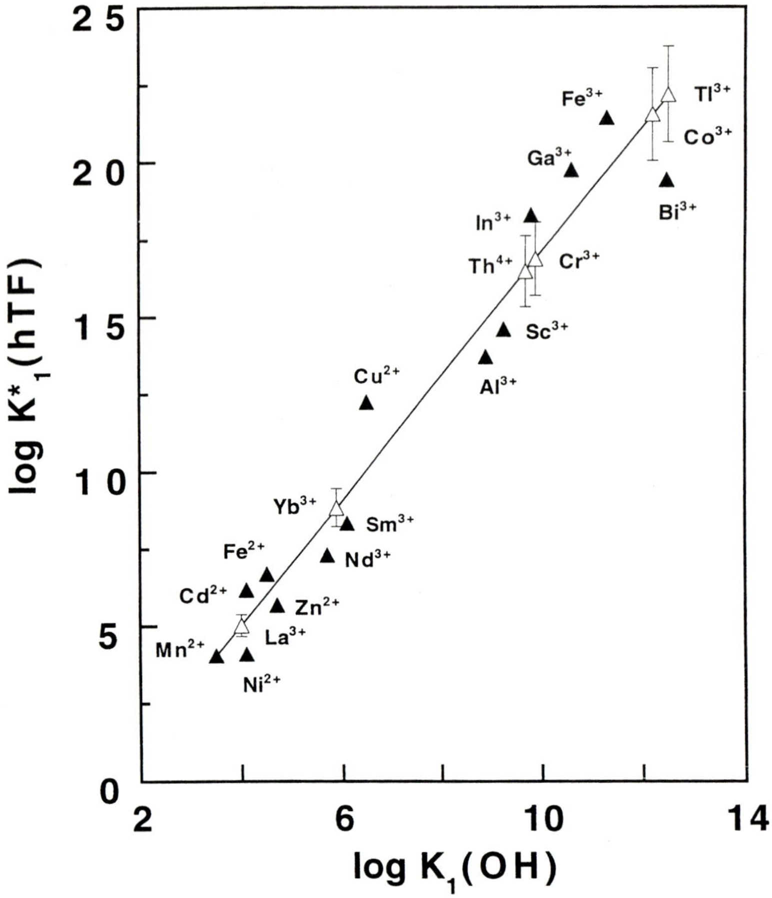

Fig. 11. Correlation of the first metal binding constant of human transferrin for metal ions with that for hydroxide binding (or metal ion acidity: log $K_1(OH) = 14 - pK_a$, where K_a is the hydrolysis constant). *Closed triangles* (▲) are for experimental data, *open triangles* (△) represent predicted values. Adapted from Ref. [72]; see Table 6 for slope and intercept

7 Rationalisation of Stability Constants for Other Proteins

Various methods have been used to predict stability constants for metal ion binding to small ligands. These include linear-free-energy relationships, acid-base methods and some empirical methods, but there are few reported attempts to make this kind of prediction for macromolecules. Figure 11 has shown that there is a good correlation between the strength of metal binding to transferrin and small ligands such as RO^-, suggesting that certain types of ligands may play a dominant role in metal binding to proteins. Such correlations of the binding constants of proteins and low molecular mass ligands for metal ions can be also extended to other proteins. The lack of previous attempts to do this [107] is due to the fact that few data on the stability constants for protein binding to a wide range of metal ions are known, which in turn is probably a reflection of the fact that such determinations cannot be made for many proteins because sufficient quantities of pure material have not been available. Since it is likely that such a bank of metal-protein stability constants will be assembled only very slowly, predictive procedures could be very important for biological, pharmacological, toxicological and other purposes.

7.1 Carbonic Anhydrase and Carboxypeptidase

These enzymes are widespread in animal and vegetable tissues as well as in bacteria. The function of carbonic anhydrase (CA) is to catalyse the hydration of carbon dioxide; that of carboxypeptidase (CP) is to catalyse the hydrolysis of certain C-terminal peptide bonds of proteins and polypeptides. Both enzymes consist of a single polypeptide chain with molecular masses of 30 kDa (CA) and 35 kDa (CP), respectively, and have a zinc ion at their active sites. The structures of the zinc binding sites are shown in Fig. 12. For CP it consists of two oxygens

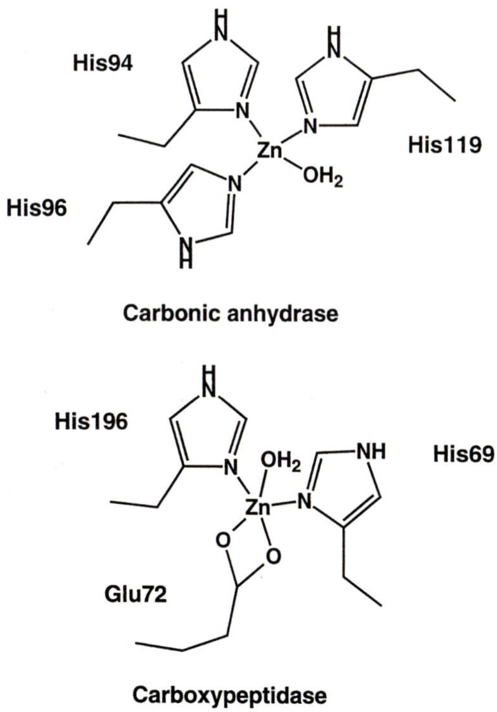

Fig. 12. Zinc binding sites in carbonic anhydrase and carboxypeptidase

from the carboxylate group of Glu-72 , two nitrogens from His-69 and 196, and oxygen from a water molecule, and for CA three imidazole groups from His-94, His-96 and His-119 and oxygen from a water molecule. Zinc can be removed from the enzymes and replaced by other metal ions, such as Hg^{2+} and Cd^{2+}, which usually results in a decrease in or loss of enzyme activity. Therefore it is important to study the thermodynamics of metal ion binding to these enzymes.

Figure 13 shows that the strength of metal binding to carbonic anhydrase [108] and carboxypeptidase [109] can be correlated with the strength of metal binding to imidazole – a nitrogen donor ligand (and part of the side-chain of His) – and is not related to the sizes of the metal ions. This suggests that nitrogens from the side-chains of the His residues are the dominant ligands in these enzymes. A previous analysis of binding constants for carboxypeptidase [107] had suggested a correlation with binding to low molecular mass N/S ligands (but not with amino acids). Since sulphur ligands were subsequently shown not to be involved in the coordination site, this serves to emphasise that such correlations may not lead to reliable predictions of stability constants when the nature of the site is unknown. Also, even when this site is known, certain metal ions may switch to alternative stronger binding sites on the protein. From Fig. 13, binding constants can be predicted for Fe^{2+} of log K_1 6.2 and 7.1, respectively, for CA and CP, and for Pb^{2+} 8.9 and 9.1, respectively. The latter ion is known to bind strongly to these enzymes [110].

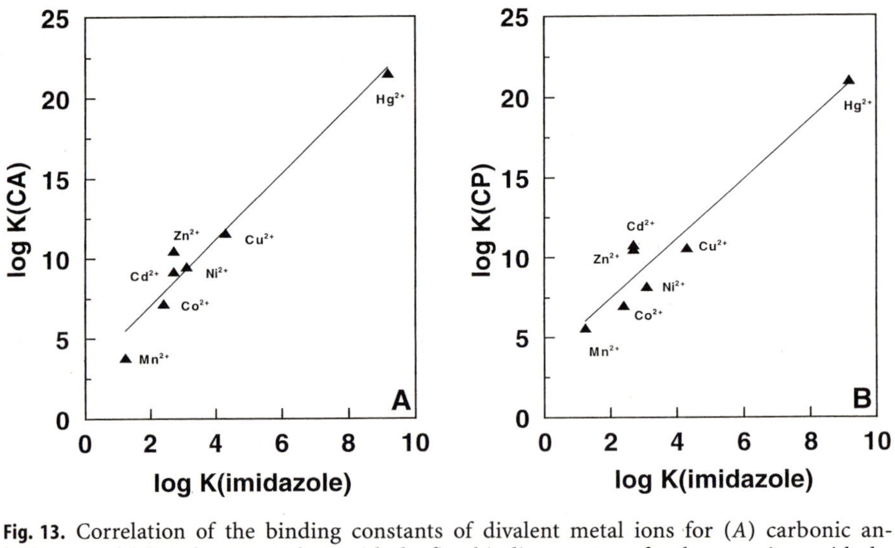

Fig. 13. Correlation of the binding constants of divalent metal ions for (*A*) carbonic anhydrase, and (*B*) carboxypeptidase, with the first binding constant for the same ions with the low molecular mass ligand imidazole [108, 109]. For slopes and intercepts see Table 6. Figure adapted from Ref. [72]

8
Preorganisation of Metal Binding Sites in Proteins

For small ligands, correlations of stability constants for binding of different metal ions to various ligands are well known, as discussed by Hancock and Martell [111–113]. Excellent linear relationships have been found between the binding of hydroxide and negatively charged low molecular mass O-donor ligands to a wide range of divalent, trivalent and quadrivalent metal ions, showing that the design of strongly-binding negatively-charged ligands can be based on metal ion acidity.

A general relationship between the logarithms of the stability constants of complexes formed by ligands with n negatively-charged oxygen atoms and log $K(OH^-)$ of the ligated metal has been derived, and the intercepts in such log-log plots [113] have been discussed in terms of Eq. 9:

$$\log K_1(\text{polydentate}) = \log K_1(OH^-) + (n\text{-}1) \log 55.5 \qquad (9)$$

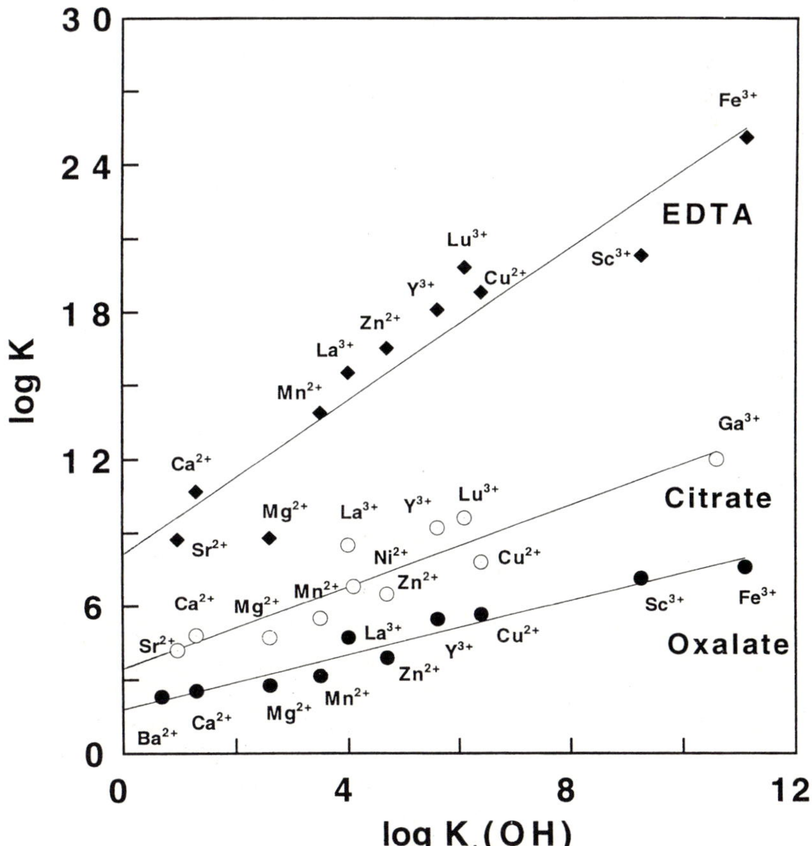

Fig. 14. Correlation of the binding constants of metal ions for hydroxide with those for bidentate (oxalate), tridentate (citrate), and hexadentate (EDTA) ligands. Data from Ref. [100]; see Table 6 for slopes and intercepts

where K_1(polydentate) refers to the stability of a complex with an *n*-dentate polydentate ligand, and 55.5 is the molality of water. The term log 55.5 represents the entropy of translation of 1 mol of water generated at a concentration of 1 molal. Therefore for bidentate ligands such as oxalate the intercept should be log 55.5 (1.74), with tridentate ligands such as citrate the intercept should be 2log 55.5 (3.49), and with hexadentate ligands such as EDTA the intercept should be 5log55.5 (8.70). The plots in Fig. 14 and data listed in Table 6 show that this is indeed the case.

The linear correlation between the strength of metal binding to transferrin and hydroxide (Fig. 11) gives an intercept of -3.4, which is much lower than expected for a tetradentate ligand (3log55.5) formed by the four protein donors: 2Tyr, 1Asp and 1His. Plots with much smaller intercepts than expected have previously been found for ligands which form large chelate rings, for example the correlation between desferrioxamine-B and hydroxide has an intercept of ca. −1.3 (instead of 5log55.5), octane-1,8-dihydroxamate an intercept of near zero (instead of 3log 55.5), and succinate of -2.3 (instead of log 55.5). These data have been interpreted in terms of the unfavourable entropy contributions associated with immobilisation of the long chelating arms of the ligands [113]. For transferrin, one possible interpretation of the negative intercept is that the structure is flexible and that after metal binding the protein conformation is more restricted with a resultant loss of entropy. Such an interpretation is in line with X-ray data: the conformation of the lobe changes from open to closed when the metal binds [4, 7]. Also, after lobe closure there is likely to be ordering of water molecules within the cleft, as found by X-ray crystallography [20]. Carbonate binding may also contribute to the entropy change, although the anion may prebind to apotransferrin before the metal enters the interdomain cleft. The interpretation of the intercept in terms of entropy effects is also supported by correlations between the strength of metal binding to the enzymes carbonic anhydrase and carboxypeptidase and binding to imidazole.

Unlike transferrin, which has a highly flexible metal binding site and is known to change its conformation on binding metal ions (the "Venus Fly-trap"), CA and CP are much more rigid and have been cited as examples of the 'entatic state' where the protein preorganises the disposition of the amino acid side-chain ligands [114]. Recent X-ray crystallographic studies on the native and apo forms of carbonic anhydrase at 1.4 Å resolution have demonstrated that they are almost identical [115] (Fig. 15) except that a water molecule close to zinc moves by about 0.8 Å. Therefore it could be said that the binding groups in the enzyme are constrained in much the same way as those on low molecular mass chelating ligands such as nitrilotriacetate (Table 6). Indeed the intercepts for the linear correlations of CA and CP binding to metal ions with that for imidazole are 2.91 and 3.72 for CA and CP, respectively, which are very close to 2log 55.5 (3.49), i.e. similar to the predicted value for a tridentate low molecular mass ligand. Carboxypeptidase is tridentate in the sense that 3 donor side-chains from the protein coordinate to Zn^{2+}, although both oxygens of the Glu residue are involved in binding. It will be interesting in future work to see if a similar behaviour also exists for other metalloproteins. This approach may be useful in the study of engineered proteins [116]. In a recent example, a 3His metal bind-

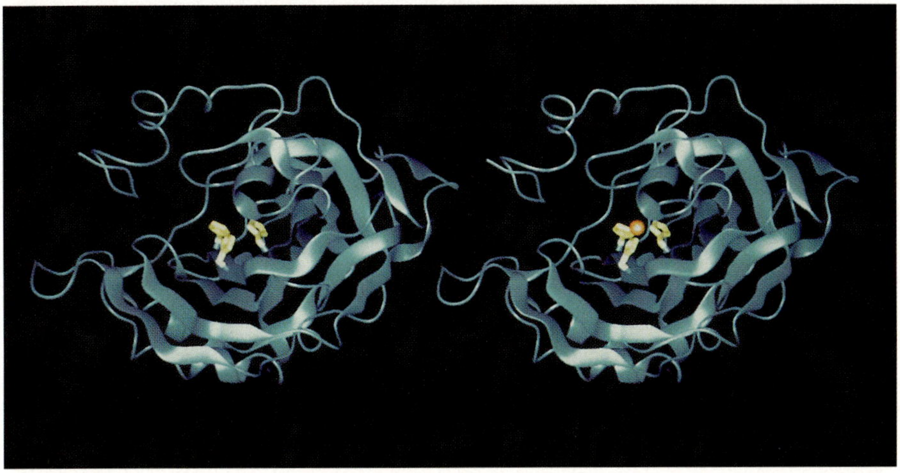

Fig. 15. Effect of metal removal on the structure of carbonic anhydrase. The X-ray structures of apo-(*left*) and holo-carbonic anhydrase (*right*) are almost identical [115]. The protein backbone is shown in blue, zinc in brown and the ligands (three His residues) in yellow. Pictures based on data from Ref. [115]

ing site similar to the carbonic anhydrase site has been engineered into the β-sheet scaffold of a 37 residue scorpion toxin [117]. A plot of log K for the limited range of metal ions Cu^{2+}, Cd^{2+}, Ni^{2+}, Zn^{2+} and Mn^{2+} versus log K (imidazole) has an intercept of 2.2 suggesting that this site is reasonably rigid.

9 Conclusion

Although the serum iron-transport protein transferrin has two very similar metal binding sites, it is clear that they differ slightly in their affinities for metal ions, as has been known for a long time [35]. There are now several methods which can be used to distinguish between these two sites, including both ^{1}H and ^{13}C-NMR spectroscopy with the use of isotopically-labelled proteins.

There is considerable interest in the strength of binding of a wide range of metal ions to human serum transferrin, on account of their natural, therapeutic, toxic and diagnostic importance. The strength of binding of metal ions cannot be rationalised on the basis of ionic size, however there is an excellent correlation with metal ion acidities. This allows predictions to be made for other metal ions for which the binding constants are not known.

Correlations of the strength of metal binding to proteins with that for binding of the same metal ions to low molecular mass ligands, such as hydroxide and imidazole, appear to provide information about entropy changes which accompany metal binding to proteins. Transferrin is a flexible metal binding protein and there is considerable loss of entropy on metal binding: the intercept in the log-log correlation with hydroxide is negative. In contrast, for the enzymes carbonic anhydrase and carboxypeptidase which have preorganised, "entatic state", metal binding sites, the intercepts are positive and close to those expected for tridentate chelating ligands. With the increasing availability of large quantities of pure recombinant metalloproteins it should be possible in future to investigate in more detail the factors which determine the thermodynamic stability of metal-protein complexes.

Acknowledgements. Our work in this area has been supported by GlaxoWellcome plc., Engineering and Physical Science Research Council, Biotechnology and Biological Sciences Research Council, and EC COST program. We thank the University of London Intercollegiate Research Service, GlaxoWellcome and Medical Research Council for the provision of NMR facilities, and are grateful to Professor R.C. Woodworth, Dr. A.B. Mason (University of Vermont, USA) and Professor R.T.A. MacGillivray (University of British Columbia, Canada) and Dr T.A. Frenkiel (National Institute of Medical Research) for collaboration. We are grateful to the Brookhaven Protein Data Bank, Dr. H.J. Zuccola (Harvard), Professor P.F. Lindley (Daresbury) and Professor E.N. Baker (Massey University) for supplying X-ray coordinates.

10 References

1. Dautry-Varsat A (1986) Biochimie 68:375
2. Octave JN, Schneider YJ, Trouet A, Crichton RR (1983) Trends Biochem Sci 8:217
3. Zuccola HJ (1993) Ph.D. Thesis, Georgia Inst Tech, Atlanta, GA, USA
4. Anderson BF, Baker HM, Norris GE, Rumball SV, Baker EN (1990) Nature (London) 344:784
5. Anderson BF, Baker HM, Dodson EJ, Norris, GE, Rumball SV, Waters JM, Baker EN (1987) Proc Natl Acad Sci USA 84:1769
6. Anderson BF, Baker HM, Norris, GE, Rice DW, Baker EN (1989) J Mol Biol 209:711
7. Haridas M, Anderson BF, Baker EN (1995) Acta Cryst D51:629
8. Kurokawa H, Mikami B, Hirose M (1995) J Mol Biol 254:196
9. Bailey S, Evans RW, Garratt RC, Gorinsky B, Hasnain SS, Horsburgh C, Jhoti H, Lindley PF, Mydin A, Sarra R, Watson JL (1988) Biochemistry 27:5804
10. Louie GV (1993) Curr Opin Struct Biol 3:401
11. Baker HM, Anderson BF, Brodie AM, Shongwe MS, Smith CA, Baker EN (1996) Biochemistry 35:9007
12. Gerstein M, Anderson BF, Norris GE, Baker EN, Lesk AM, Chothia C (1993) J Mol Biol 234:357
13. Grossmann JG, Neu M, Evans RW, Lindley PF, Appel H, Hasnain SS (1993) J Mol Biol 229:585
14. Grossmann JG, Neu M, Pantos E, Schwab FJ, Evans RW, Townes-Andrews E, Lindley PF, Appel H , Thies W-G, Hasnain SS (1992) J Mol Biol 225:811
15. Castellano AC, Barteri M, Bianconi A, Borghi E, Castagnola L, Castagnola M, Della Longa S, La Monaca A (1993) Biophys J 64:520; Castellano AC, Barteri M, Castagnola M, Bianconi A, Borghi E, Della Longa S (1994) Biochem Biophys Res Comm 198:646
16. Dewan JC, Mikami B, Hirose M, Sacchettini JC (1993) Biochemistry 32:11963
17. Grossmann JU, Mason AB, Woodworth RC, Neu M, Lindley PF, Hasnain SS (1993) J Mol Biol 231:554
18. Faber HR, Bland T, Day CL, Norris GE, Tweedie JW, Baker EN (1996) J Mol Biol 256:352
19. Egan TJ, Zak O, Aisen P (1993) Biochemistry 32:8162
20 Baker EN (1994) Adv Inorg Chem 41:389
21. Harris WR (1986) J Inorg Biochem 27:41
22. Li H, Sadler PJ, Sun H (1996) J Biol Chem 271:9483
23. Harris WR (1986) Inorg Chem 25:2401
24. O´Hara PB, Koenig SH (1986) Biochemistry 25:1445
25. Zak O, Aisen P (1988) Biochemistry 27:1075
26. Ward SG, Taylor RC (1988) In: Gielen MF (ed) Metal-Based Anti-Tumour Drugs, Freund Publishing House Ltd, London, pp1-54
27. Harris WR, Sheldon J (1990) Inorg Chem 29:119

28. Kratz F, Hartmann M, Keppler BK, Messori L (1994) J Biol Chem 269:2581
29. Smith CA, Sutherland-Smith AJ, Keppler BK, Kratz F, Baker EN (1996) J Biol Inorg Chem 1:424
30. Lehmann M, Culig H, Taylor DM (1983) Int J Radiat Biol 44:65
31. Ford-Hutchinson AW, Perkins DJ (1971) Eur J Biochem 21:55
32. Cox MC, Barnham KJ, Sadler PJ, unpublished results
33. Schlabach MR, Bates GW (1975) J Biol Chem 250:2182
34. Aisen P, Aasa R, Redfield AG (1969) J Biol Chem 244:4628
35. Harris DC, Aisen P (1989) In: Loehr TM (ed) Iron Carriers and Iron Proteins, VCH, Weinheim, Germany, pp239
36. Zweier JL, Peisach J, Mims WB (1982) J Biol Chem 257:10314
37. Garratt RC, Evans RW, Hasnain SS, Lindley PF (1986) Biochem J 233:479
38. Hasnain SS, Evans RW, Garratt RC, Lindley PF (1987) Biochem J 247:369
39. Garratt RC, Evans RW, Hasnain SS, Lindley PF (1991) Biochem J 280:151
40. Mangani S, Messori L (1992) J Inorg Biochem 46:1
41. Smith CA, Anderson BF, Baker HM, Baker EN (1992) Biochemistry 31:4527
42. Smith CA, Anderson BF, Baker HM, Baker EN (1994) Acta Cryst D50:302
43. Shongwe MS, Smith CA, Ainscough EW, Baker HM, Brodie AM, Baker EN (1992) Biochemistry 31:4451
44. Day CL, Anderson BF, Tweedie JW, Baker EN (1993) J Mol Biol 232:1084
45. Lindley PF, Bajaj M, Evans RW, Garratt RC, Hasnain SS, Jhoti H, Kuser P, Neu M, Patel K, Sarra R, Strange R, Walton A (1993) Acta Cryst D49:292
46. Jhoti H, Gorinsky B, Garratt RC, Lindley PF, Walton AR, Evans RW (1988) J Mol Biol 200:423
47. Makey DG, Seal US (1976) Biochim Biophys Acta 453:250
48. Princiotto JV, Zapolski EJ (1975) Nature (London) 255:87
49. Zapolskil EJ, Princiotto JV (1980) Biochemistry 19:3599
50. Lin L, Mason AB, Woodworth RC, Brandts JF (1993) Biochemistry 32:9398
51. Lin L, Mason AB, Woodworth RC, Brandts JF (1994) Biochemistry 33:1881
52. Cannon JC, Chasteen ND (1975) Biochemistry 21:4573
53. Zweier JL, Wooten JB, Cohen JS (1985) Biochemistry 20:3505
54. Harris DC (1977) Biochemistry 16:560
55. Kubal G, Sadler PJ, Tucker A (1994) Eur J Biochem 220:781
56. Mason AB, Miller MK, Funk WD, Banfield DK, Savage KJ, Oliver RWA, Green BN, MacGillivray RTA, Woodworth RC (1993) Biochemistry 32:5472
57. Kubal G, Mason AB, Patel SU, Sadler PJ, Tucker A, Woodworth RC (1993) Biochemistry 32: 3387
58. Kubal G, Mason AB, Sadler PJ, Tucker A, Woodworth RC (1992) Biochem J 285:711
59. Beatty EJ (1995) PhD thesis, University of London
60. Aramini JM, Krygsman PH, Vogel HJ (1994) Biochemistry 33:3304
61. Aramini JM, Vogel HJ (1993) J Am Chem Soc 115:245
62. Aramini JM, Vogel HJ (1994) J Am Chem Soc 116:1988
63. Germann MW, Aramini JM, Vogel HJ (1994) J Am Chem Soc 116: 6971; Aramini JM, McIntyre DD, Vogel HJ (1994) J Am Chem Soc 116:11506
64. Saponja JA, Vogel HJ (1996) J Inorg Biochem 62:253; Bulter A, Eckert H (1989) J Am Chem Soc 111:2802
65. Kiang W, Sadler PJ, Reid DG (1993) Magn Reson Chem 31:S110
66. Aramini JM, Vogel HJ (1996) J Magn Reson B110:182
67. Weslund PO, Wennerstrom H (1982) J Magn Reson 50:451
68. Werbelow LG (1979) J Chem Phys 70:5381; Werbelow LG, Pouzard G (1981) J Phys Chem 85:3887
69. Beatty EJ, Cox MC, Frenkiel TA, Tam BM, Mason AB, MacGillivray RTA, Sadler PJ, Woodworth RC (1996) Biochemistry 35:7635
70. Beatty EJ, Cox MC, Frenkiel TA, Kubal G, Mason AB, Sadler PJ, Woodworth RC (1994) In: Collery PH, Littlefield NA, Etienne JC (eds) Metal Ions in Biology and Medicine, John Libby Eurotext, Paris, p315

71. Cox MC, Li H, Mason AB, Sadler PJ, Sun H, Woodworth RC, unpublished results
72. Li H, Sadler PJ, Sun H (1996) Eur J Biochem 242:387
73. Harris WR, Chen Y (1994) J Inorg Biochem 54:1
74. Aasa R, Malmstrom B, Saltman P, Vänngard T (1963) Biochim Biophys Acta 75:203
75. Aisen P, Leibman A, Reich HA (1966) J Biol Chem 241:1666
76. Binford JS, Jr, Foster JC (1974) J Biol Chem 243:407
77. Fletcher J, Huehns ER (1967) Nature (London) 215:584
78. Fletcher J, Huehns, ER (1968) Nature (London) 218:1211
79. Harris DC, Aisen P (1975) Biochemistry 14:262
80. Aisen P, Leibman A (1968) Biochem Biophys Res Comm 30:407
81. Evans RW, Williams J (1978) Biochem J 173:543
82. Aisen P, Leibman A, Zweier J (1978) J Biol Chem 253:1930
83. Harris WR, Chen Y, Wein K (1994) Inorg Chem 33:4991
84. Princiotto JV, Zapolski EJ (1975) Nature (London) 255:87
85. Lestas AN (1976) Br J Haematol 32:341
86. Baldwin DA, de Sousa DMR, von Wandrszka RMA (1982) Biochim Biophys Acta 719:140
87. Bali, PK, Aisen, P (1992) Biochemistry 31:3963
88. Harris WR, Stenback JZ (1988) J Inorg Biochem 33:211
89. Chasteen ND, Williams J (1981) Biochem J 193:717
90. Williams J, Chasteen ND, Morton K (1982) Biochem J 201:527
91. Folajtar DA, Chasteen ND (1982) J Am Chem Soc 104:5775
92. Williams J (1982) Biochem J 201:647
93. Zweier JL (1978) J Biol Chem 253:7616
94. Zweier JL, Aisen P (1977) J Biol Chem 252:6090
95. Shannon RD (1976) Acta Crystallogr 32A:751
96. Smith CA, Ainscough EW, Baker HM, Brodie AM, Baker EN (1994) J Am Chem Soc 116: 7889
97. Baxter GF (1989) Pharm J 243:805
98. Harris WR, Pecoraro VL (1983) Biochemistry 22:292
99. Harris WR (1989) Adv Exp Med Biol 249:67
100. Pettit G, Pettit LG (1993) IUPAC Stability Constants Database, IUPAC/Academic Software, Otley, United Kingdom
101. Hirose J, Fujiwara H, Magarifuchi T, Iguti Y, Iwamoto H, Kominami S, Hiromi K (1996) Biochim Biophys Acta 1296:103
102. Bertini I, Messori L, Pellacani GC, Sola M (1988) Inorg Chem 27:761
103. Bertini I, Luchinat, C, Messori L (1983) J Am Chem Soc 105:1347
104. He Q, Mason AB, Woodworth RC (1996) Biochem J 318:145
105. Harris WR, Carrano CJ, Pecoraro VL, Raymond KN (1981) J Am Chem Soc 103:2231
106. Pecoraro VL, Harris WR, Carrano CJ, Raymond KN (1981) Biochemistry 20:7033
107. Vallee BL, Williams RJP, Coleman JE (1961) Nature (London) 190:633
108. Lindskog S, Nyman PO (1964) Biochim Biophys Acta 85:462
109. Coleman JE, Vallee BL (1961) J Biol Chem 236:2244
110. Lindskog S (1970) Struct Bond 8:153
111. Dimmock PW, Warwick P, Robbins RA (1995) Analyst 120:2159
112. Evers A, Hancock RD, Martell AE, Motekaitis RJ (1989) Inorg Chem 28:2189
113. Hancock RD, Martell AE (1986) Chem Rev 89:1875
114. Vallee BL, Williams RJP (1968) Proc Natl Acad Sci USA 59:498
115. Håkansson K, Carsson M, Svensson LA, Liljas A (1992) J Mol Biol 227:1192
116. Glusker JP (1991) Adv Protein Chem 42:1
117. Vita C, Roumestand C, Toma F, Ménez A (1995) Proc Natl Acad Sci USA 92:6404

Metal Centres of Bacterioferritins or Non-Haem-Iron-Containing Cytochromes b_{557}

Nick E. Le Brun, Andrew J. Thomson and Geoffrey R. Moore*

Centre for Metalloprotein Spectroscopy & Biology, School of Chemical Sciences, University of East Anglia, Norwich NR4 7TJ, UK.* *E-mail: g.moore@uea.ac.uk*

Structural, spectroscopic and chemical features of the metal centres of non-haem-iron-containing cytochrome b_{557} are described. This haemoprotein is a member of the ferritin family of proteins and thus is also called bacterioferritin. It consists of 24 polypeptide subunits and, in addition to its inter-subunit bis-methionine coordinated haem b groups, it also contains intra-subunit dinuclear metal centres and has the capacity for a central non-haem-iron deposit of 4,500 Fe(III) ions per molecule. The dinuclear site can be occupied by Co(II) or Mn(II), and probably by Zn(II) and Fe(II) as well. Mixed metal centres also appear to be formed. The non-haem-iron core can be either amorphous, as customarily found with native proteins, or crystalline. Crystalline cores can be laid down in vitro. This latter type of core is often superparamagnetic and this phenomenon is decribed, particularly with regards to its characterisation by ^{57}Fe Mössbauer spectroscopy, EPR spectroscopy and Magnetic circular dichroism spectroscopy. Finally, possible functional roles for the metal centres are described.

Keywords: Ferritin, bacterioferritin, cytochrome b_{557}, dinuclear, haem

* Author to whom correspondence should be addressed.

Structure and Bonding, Vol. 88

1 Introduction

Non-haem-iron-containing cytochrome b_{557}, or bacterioferritin, is a member of the ferritin family of proteins [1–5]. These are characterised by their ability to accumulate deposits of non-haem iron within a protein shell composed of 24 subunits [5–8]. Such proteins have been found in animals, plants and micro-organisms, as summarised in Table 1. The physiological functions of these proteins have been extensively investigated but in many cases their exact roles have not been established. If the term "ferritin" is taken to indicate a mobile iron store, then bacterioferritin should not be classified as a ferritin yet since it has not been demonstrated unequivocally to be a utilisable iron store, and it is not tightly regulated by iron in the way that a general iron store is expected to be. For example, BFR is produced in *Rhodobacter capsulatus* under both iron-deficient and iron-sufficient conditions in significant amounts [20], while ferritin synthesis in animals is strongly induced by iron [7, 8 and references therein]. The original name by which bacterioferritin (BFR) was known, non-haem-iron-containing cytochrome b_{557}, both acknowledges that it may not be a classical ferritin and recognises that it contains haem. It may well be [29] the same cytochrome as that described in 1964 by Deeb & Hager [30], who called it cytochrome b_1, and in 1971 by Bartsch et al. [31], who called it cytochrome b_{557}.

Amongst the proteins of Table 1, BFR is unique in possessing intrinsic haem and this has generated considerable interest in possible physiological roles in addition to iron storage. Proposed functions include a role in respiration, by virtue of its proposed identification as cytochrome b_1 [29], and as an electron store. However, respiratory cytochromes tend to be located in the periplasmic space of bacterial cells, or are transmembraneous, and BFR is a cytoplasmic protein [20], consistent with its lack of a signal sequence [32, 33]. Also, ^{57}Fe Mössbauer spectra of intact cells of *P. aeruginosa* [34] and *R. capsulatus* [20], grown either aerobically or anaerobically, indicate that the non-haem-iron storage deposits are largely in the Fe(III) state. Hence it seems unlikely that BFR has a function as an electron store during the stationary phase of growth in batch cultures. The high phosphate content of the native cores (Table 1) suggests that BFR may be involved in phosphate metabolism, but there is no further evidence to support this and, in any case, it is not clear how possession of a haem would help with this. The recent discovery of other multimeric protein families that have amino acid sequence similarities with BFR has raised the possibility that BFR has a protective role in conditions leading to oxidative stress. A particularly interesting case is that of a DNA-binding protein from starved cells of *E. coli* (Dps) [35]. This binds to DNA when the cells are starved and appears to provide some protection from the effects of H_2O_2. More recently, Peña & Bullerjahn [36] have shown that Dps from *Synechococcus* PCC7942 contains non-covalently bound haem and has weak catalase activity in vitro. Though the published amino acid sequences of Dps reveal that the haem ligands must be different from the bis-Met ligation of BFR (see Section 4), they do show that there is as much similarity between BFR and Dps as there is between BFR and FTN [33, 36, 37]. Thus further structural characterization of Dps is keenly awaited.

Our long-standing interests in haemoproteins [38, 39] led us to investigate the haem groups of BFR (see Section 4) and this in turn led onto studies of its other metal sites (see Sections 5–8). In the present article we shall describe the metal sites of ferritins with BFR as the main focus of interest. Comparison with other ferritins will only be made where there are points of particular structural, spectroscopic or mechanistic interest.

2 Functional Assays of Bacterioferritin

Though our main emphasis is on structural features of BFR, we intend to put the metal centres into a functional context where possible. This is particularly interesting when protein engineering methods have been used in concert with structural investigations. However, since the precise biological functions of BFR are not known, construction of physiologically relevant assays may not have been achieved. It is assumed that BFR is a utilisable iron store in which excess iron is deposited and can be released when there is an intracellular demand for it. However, whilst the former role is established by the isolation of iron-laden BFR, the latter has not been demonstrated. An additional complication is that there are, presumably, physiological iron donor and iron acceptor molecules and, perhaps, physiological redox partners for BFR [40]. Such species have not been identified, though it is possible that a recently characterised Fe_2S_2 ferredoxin, whose gene is contiguous with that of BFR, at least in *E. coli* [32], is one of the partners for BFR [41]. Given this background it is probable that most of

Table 1. Characteristics of ferritins

Ferritin	Sources	Number of subunits [a]	Metal centres in polypeptide [b]
animal ferritin	e.g. human, horse, rat, tadpole & limpet	At least 2: L and H chains [d]	dinuclear metal binding site in each H-chain?
plant ferritin (phytoferritin)	e.g. pea, clover & soybean	Possibly one basic type but extensions to it exist [e]	not determined
Bacterial ferritin (FTN) [f]	Many bacteria e.g. *E. coli* & *Helicobacter pylori*	Not certain only single chain recombinant proteins studied in detail	dinuclear Fe centre and associated mononuclear Fe centre in each subunit
Bacterioferritin (BFR) or non-haem-iron-containing cytochrome b_{557} [g].	Many bacteria e.g. *E. coli*, *Pseudomonas aeruginosa* & *Azotobacter vinelandii*	Some have one type of subunit and others (e.g. *P. aeruginosa*) have two: α & β	dinuclear metal centre in each subunit and 12 intersubunit bis-Met coordinated haem binding sites

[a] Multiple subunit types have been identified for most families of ferritins. The widely held view is that cell-specific heteropolymers of different subunit types exist, at least for L and H chain ferritins, for which there is some experimental support [9, 10]. To date, however, all ferritins prepared using recombinant DNA procedures are homopolymers.

[b] Only metal centres unambiguously defined by spectroscopic or crystallographic methods are included in the table. These are described in Sections 4 (haem groups), 5 (intra-subunit di-nuclear metal sites) and 7 (non-haem-iron cores). Intra-subunit di-nuclear metal centres in H-chain ferritin are not as well characterised as those of the other proteins indicated. Crystallographic studies show that metal ions such as Tb(III) do bind at the site and kinetic and ^{57}Fe Mössbauer studies indicate that di-iron centres can be formed [reviewed in 8] but there is no direct crystallographic evidence for di-iron centres in H-chain ferritin.

[c] All values given in the table are for samples of wild-type proteins as isolated using non-recombinant DNA procedures. Such holoproteins are usually polydisperse samples with variable core sizes. Various studies of samples with cores laid down in vitro have been reported, such as that for *E. coli* FTN which has a T_B of 19 K for cores produced at a Fe:P ratio of 4:1 [18]. The Mössbauer data are generally taken to be an indicator of core morphology (see Section 7.4). When $T_{ord} > T_B$ and T_B is relatively high the core has a high degree of crystallinity.

Characteristics of the non-haem-iron (NHI) cores [c]				References
NHI/molecule	P/molecule	Fe/P ratio	Mössbauer T_B	
1000 – 3000	variable, but low	≥ 10:1	T_{ord}> T_B and T_B 40–50 K.	5, 6, 8 – 11 & references therein
1800 1300	640 325	2.8:1 4:1	not determined	7, 12 – 15
770 – 1570	600 – 1000	1.3:1 – 1.6:1	not determined	8, 16 – 19
600 – 2300	320 – 1500	1.1:1 – 1.9:1	T_B > T_{ord} and T_B >3K	1 – 3, 11, 16, 20 – 27

[d] Other forms of subunit type exist. For example, frog ferritin has an M or H' subunit as well as L and H subunits [7].

[e] Plant ferritins have an N-terminal transit peptide required for directing the newly synthesized protein to the correct cellular location and an N-terminal extension of 28 amino acids compared to other ferritins [12]. The transit peptide is not attached to subunits in the mature iron-laden protein but the extension peptide may be.

[f] FTN was initially called prokaryotic ferritin or PFR [16]

[g] The name "haemoferritin" was introduced to describe horse spleen ferritin with haem bound to it at a low-spin site [67, 68] (see Section 4.2). However, the name"haemoferritin" is now used by some researchers to describe bacterioferritin [e.g. 8].

the in vitro assays of BFR function are non-physiological. We shall briefly describe two assays which have been employed to study the roles of the metal centres of BFR.

2.1 The Aerobic Oxidative Iron Uptake Assay

This assay is sometimes referred to as the ferroxidase assay [e.g. 42]. In the presence of air a sample of suitably buffered BFR is placed into a cuvette and an aliquot of an Fe(II) solution added. The time course of oxidation of the added Fe(II) is monitored by measuring the absorbance increase at 340–410 nm which is associated with formation of the brown Fe(III) colour. Typical time-course profiles are shown in Fig. 1. The rate of oxidation of Fe(II) in the presence of BFR is considerably higher than in its absence. Thus, in vitro, BFR, like mammalian and plant ferritins [8 and references therein, 14, 15], catalyses the oxidation of Fe(II) by O_2. Close inspection of Fig. 1 shows that there is an initial jump in absorption at 340 nm before the rate of change of absorbance decreases. This initial jump has been termed phase 2 of the reaction, phase 1 being a reversible haem Soret band perturbation that takes place extremely rapidly without either the ferrihaem or Fe(II) changing their redox states [44]. Phase 3 of the reaction is the slower build-up of the 340 nm absorption. Further mechanistic analyses of these reaction phases are deferred until see Section 8, but we note now that mutation of residues at the intra-subunit dinuclear metal binding centre (see Section 5) changes both the phase 2 and phase 3 time course profiles (Fig. 1).

An alternative method for monitoring the ferroxidase activity of ferritins has been reported by Chasteen and co-workers [45, 46]. In their approach, a rapid-response oxygen microelectrode is used to monitor the rate of O_2 consumption as Fe(II) is oxidised. With the addition of 24 Fe(II) ions per ferritin 24mer H_2O_2 is the product of O_2 reduction according to equation (1) but with larger additions (240–960 Fe(II) ions per 24mer) the Fe(II):O_2 stoichiometry increases to that given by equation (2). Reaction (1) is proposed to occur at dinuclear sites in the protein shell (see Section 5) with reaction (2) taking place on the growing mineral core. Unfortunately, this method does not seem to have been applied to BFR.

$$2Fe^{2+} + O_2 + 4H_2O \rightarrow 2FeOOH_{core} + H_2O_2 + 4H^+ \tag{1}$$

$$4Fe^{2+} + O_2 + 6H_2O \rightarrow 4FeOOH_{core} + 8H^+ \tag{2}$$

2.2 The Reductive Iron Release Assay

A commonly used iron release assay employs an Fe(II) chelator to bind to Fe(II) released from BFR, creating a coloured species that can be monitored spectrophotometrically. Ferrozine is now the ligand of choice [47]. Appreciable release of iron does not result unless the core iron is reduced. This is achieved by the addition of a reductant which, in reported work with BFR, is generally a quinol

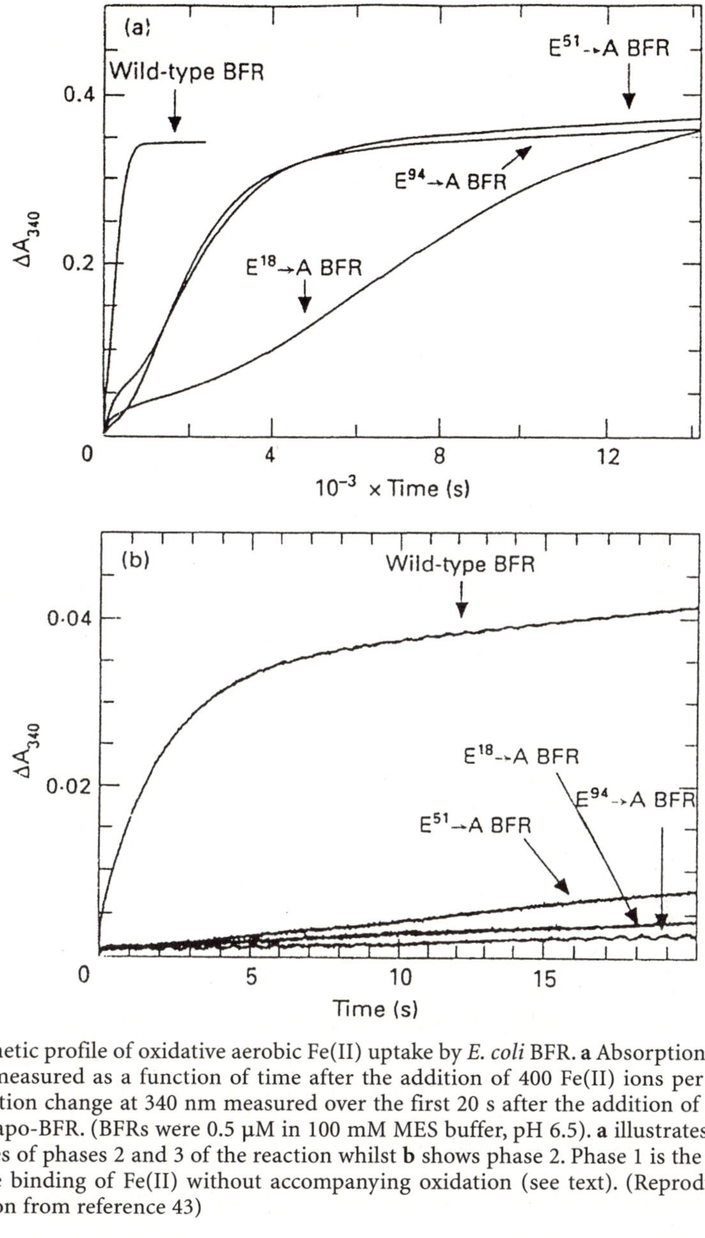

Fig. 1. Kinetic profile of oxidative aerobic Fe(II) uptake by *E. coli* BFR. **a** Absorption change at 340 nm measured as a function of time after the addition of 400 Fe(II) ions per apo-BFR. **b** Absorption change at 340 nm measured over the first 20 s after the addition of 400 Fe(II) ions per apo-BFR. (BFRs were 0.5 μM in 100 mM MES buffer, pH 6.5). **a** illustrates the overall profiles of phases 2 and 3 of the reaction whilst **b** shows phase 2. Phase 1 is the rapid and reversible binding of Fe(II) without accompanying oxidation (see text). (Reproduced with permission from reference 43)

species [40], though flavins and other reductants have been used in similar assays [e.g. 48, 49]. Figure 2 illustrates the main features of the assay.

Considering just the iron species, the general reaction scheme is:

slow $$\text{Core} - Fe^{3+}_{n} + e^{-} \rightarrow \text{Core} - Fe^{3+}_{(n-1)} + Fe^{2+} \quad (3)$$

fast $$Fe^{2+} + \text{xligand} \rightarrow Fe^{2+}(\text{ligand})_{x} \quad (4)$$

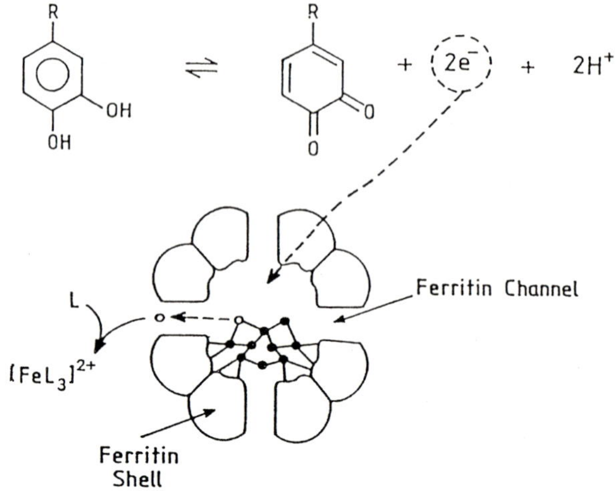

Fig. 2. Schematic view of a reductive release assay. BFR is shown in cross-section with four of the channels through the protein shell. The Fe(III) ions in the core are represented by• ; for clarity the oxide, hydroxide and phosphate in the core are omitted as well as the intra-subunit metal binding sites and the inter-subunit haem groups. Oxidation of the diphenol produces electrons and protons which cause the reductive dissolution of the core. The electrons may travel to the core via the haem groups. Fe(II) ions, represented by o, migrate through the protein shell, probably through the channels, and bind to the ligand (L) at the surface of the molecule or in the bulk phase

Provided reaction (4) is much faster than reaction (3) the overall measured rate corresponds to the rate of reaction (3). Though equations (3) and (4) are sufficient to allow the rate of iron released from ferritins by a reductive process to be determined [40, 47], they do not adequately describe the chemistry that is occurring. Thus, there is no indication of what happens to the core phosphate, nor of the involvement of protons. It is clear that the reduction of the oxide-rich ferritin cores is a heavily proton-consuming reaction, as it has a stoichiometry of $\sim 2H^+$ per electron added [50]. *A. vinelandii* BFR, however, appears to have a lower requirement for protons [25].

3
Protein Structure of Bacterioferritin

The X-ray structure of *E. coli* apo-BFR [21, 22] reveals that it strongly resembles the structures of animal ferritins [8, 51, 52], as anticipated by amino acid sequence comparisons and molecular modelling studies [e.g 4, 53]. It is an almost spherical coat of protein surrounding a central cavity (Fig. 3). The protein coat is ~20 Å thick and is composed of 24 identical subunits packed together in a highly symmetrical form. The central cavity has a diameter of ~80 Å and is where the non-haem-iron polynuclear clusters are located in the holo-BFR. The overall structure exhibits 432 symmetry in which the subunits are related by operation of 4-fold, 3-fold and 2-fold symmetry axes [8, 22]. Subunit packing at both the eight 3-fold and six 4-fold axes is such that narrow channels are formed that run through the protein coat linking the central cavity with the outside of the molecule (Fig. 3). In BFR both types of channel are hydrophilic, unlike in human ferritin in which only the 3-fold channels are hydrophilic [8], and both are sufficiently wide to allow passage of water molecules. There has been extensive investigation of the possible roles of the channels in animal ferritin [reviewed in 8] and it seems probable that the 3-fold channels provide the pathway for iron uptake into the central cavity [e.g. 54–56]. There have only been limited studies reported concerning the role of the 3-fold channels of BFR. Le Brun et al. [43] showed that the Asp118Ala variant of *E. coli* BFR took up Fe(II) in the oxidation assay of Fig. 1 with rates virtually unchanged from that of wild-type BFR. In contrast, variants of human H-chain ferritin with similar 3-fold channel mutations, Asp131Ala and Glu134Ala, had slower rates of iron uptake than the wild-type protein [55]. Further studies of BFR are required, however.

Each subunit consists of an antiparallel 4-α-helical bundle with a short fifth α-helix almost orthogonal to the bundle at the C-terminal end. Each subunit is ~50 Å long with a transverse diameter across the 4–helical bundle of ~25 Å. The 2-fold dimer interface formed by the packing of two subunits is also a 4-α-helical bundle. Further details of these structural features for animal ferritins and BFR can be found in Harrison & Arosio [8] and Trikha et al. [52], and Frolow et al. [22], respectively.

As well as the metal centres that are the main focus of this article, there is experimental support for a fluorescent group, most probably a quinone, being present in BFR and other ferritins [3, 57]. However, it is possible that the presence of such groups is a result of oxidative damage as mass spectra of freshly prepared BFR show the protein has the mass expected from the amino acid sequence [33, 58].

4 Haem Binding to Bacterioferritin

4.1 The Intrinsic Haem

The X-ray structure of *E. coli* BFR shows that the intrinsic haem groups are located at inter-subunit sites in which the Met52 residues of adjacent subunits provide the two axial ligands to the iron (Fig. 3). Prior to determination of the X-ray structure, spectroscopic studies had indicated bis-methionine ligation, most reliably with near-infra-red MCD (Fig. 4) [23] and EXAFS [59] spectroscopy, and site-directed mutagenesis carried out concurrently with the X-ray structure determination showed that Met52 was the ligand [58]. Interestingly, Met52 is not conserved amongst all BFRs, though in sequences lacking Met52, such as those from *P. aeruginosa* [60] and *Synechocystis* PCC6803 [24] BFRs, a methionine at position 48 is present. This may provide the axial ligands since EPR and MCD spectroscopic measurements of *P. aeruginosa* BFR [61] make it

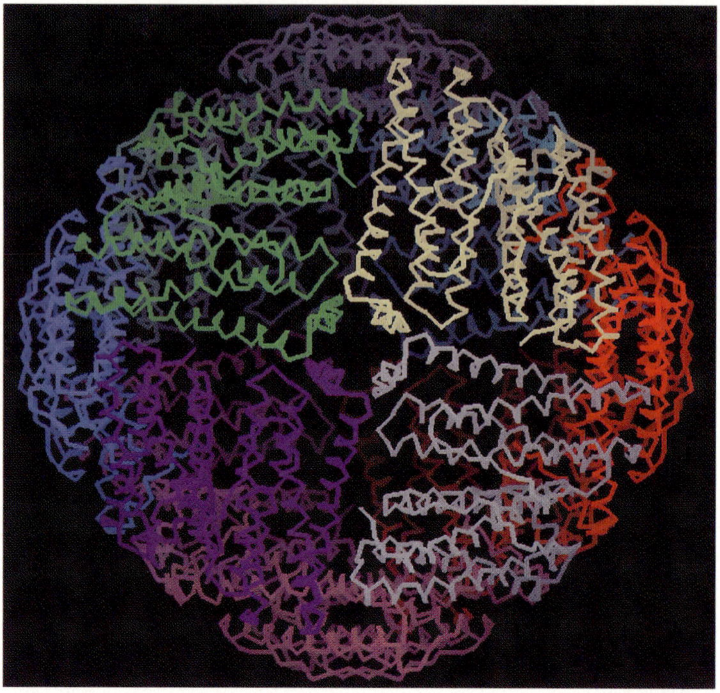

Fig. 3 a

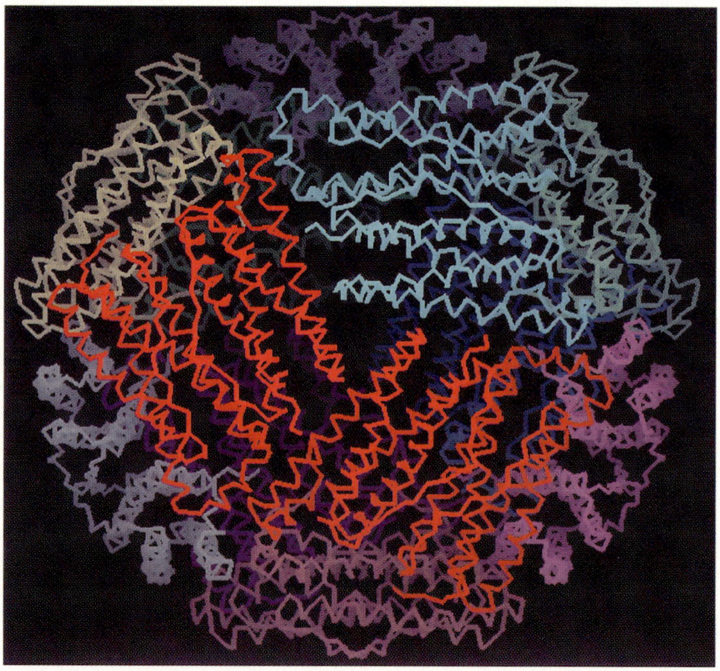

Fig. 3 b

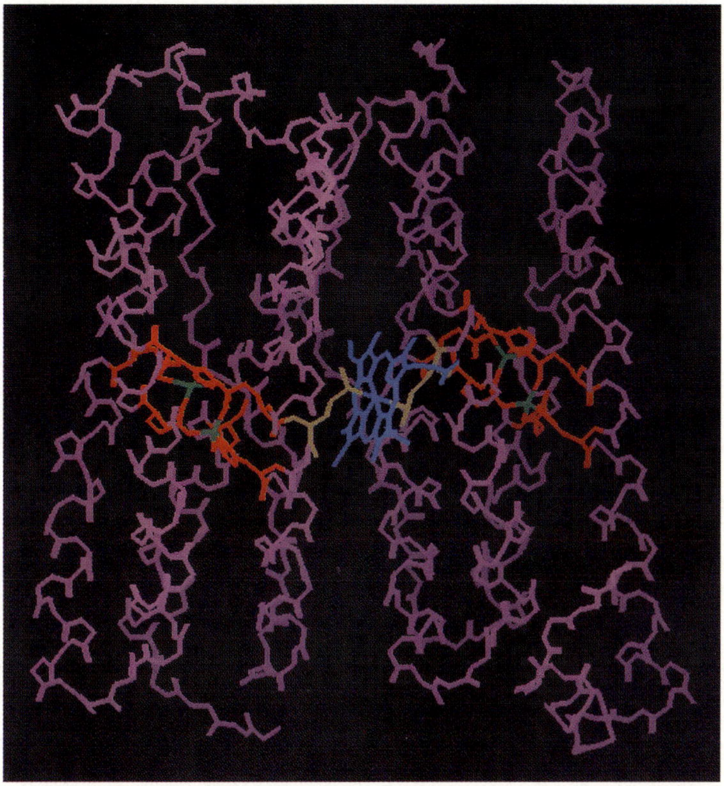

Fig. 3c

Fig. 3. The structure of *E. coli* BFR. The pictures were generated using Quanta (Molecular Simulations Inc) with the structure coordinates deposited in the Brookhaven Protein Data Bank by Frolow et al. [22]. **a** View of the surface of the 24mer with subunit dimers coloured differently to show a 4-fold channel; **b** view of the surface of the 24mer with subunit dimers coloured differently to show a 3-fold channel; **c** a subunit dimer viewed from the inner surface showing the haem group in its inter-subunit site coordinated by two methionines and the two intra-subunit di-nuclear metal ion centres with the amino acid ligands to the metal ions

clear that this protein also has bis-methionine ligation. Residues 48 and 52 are both part of helix B and their side-chains are on the same face of the helix. Thus a displacement of the helix relative to the haem should be sufficient to maintain the haem environment.

The bis-thioether axial ligation generates a strong ligand field that causes the iron to be low-spin in both its Fe(II) and Fe(III) states, and makes it inert to reaction with potential ligands such as O_2 and CN^-. The NIR charge transfer band diagnostic of bis-methionine axial ligation has a λ_{max} of 2100–2200 nm (Fig. 4). A band at this wavelength has been observed only for ferrihaems with bis-thioether ligation, such as the bis(tetrahydrothiophene) complex of Fe(III)-octaethylporphyrin [62] and the bis-methionine coordinated haem of the Arg98Cys/His102Met variant of *E. coli* b_{562} in which the haem is covalently

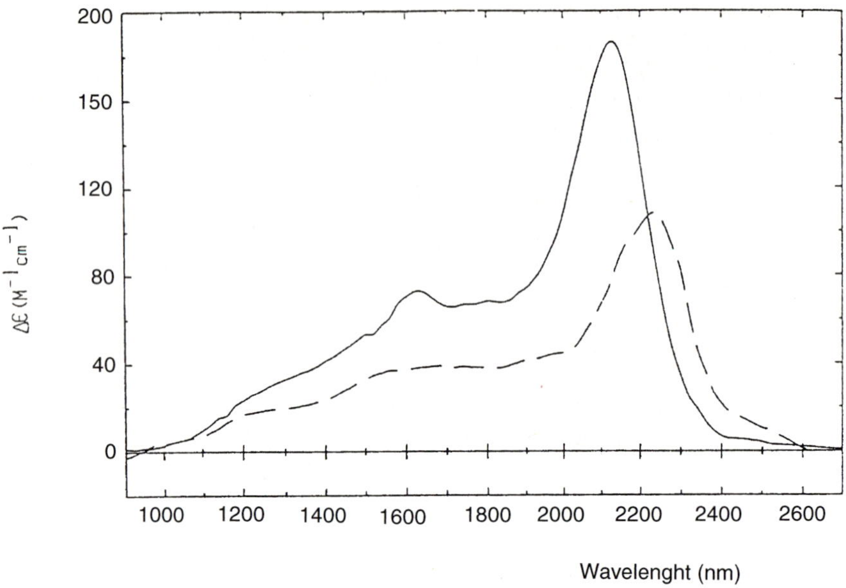

Fig. 4. The MCD spectra of *P. aeruginosa* BFR (*solid line*) and bis(tetrahydrothiophene)-Fe(III)-octaethylporphin at 5 T and 4.2 K. $\Delta\varepsilon = \varepsilon_L - \varepsilon_R$, where ε_L and ε_R are the molar extinction coefficients for left and right circularly polarised light, respectively. [The BFR concentration was 280 μM in haem in 200 μM HEPES buffer, pH 7.4, in D_2O with 50%(v/v) deuterated glycerol added; and the porphin complex was dissolved in a 2:1 (v/v) mixture of $CH_2Cl_2/(C_2H_5)_2O$)]. (Adapted from reference 39)

bound to the protein via one thioether link [63]. As with other low-spin haems, the EPR g-values are not uniquely diagnostic of bis-methionine ligation. For BFR these are $g_z = 2.88$, $g_y = 2.31$ and $g_x = 1.45$, well within the range expected for histidine-methionine ligation [38, 39].

Replacement of Met52 with His52, by site-directed mutagenesis, results in total loss of haem rather than the generation of a bis-histidine-ligated cytochrome. Although the Met52His BFR is haem-free, its activity in oxidative Fe(II) uptake assays is largely unaffected, as is its ability to form intra-subunit dinuclear metal complexes [58]. The inability of haem to bind in the bis-histidine pocket is probably related to the different sizes and shapes of histidine and methionine, which would lead the histidine ring Ns to be further away than methionine S from the haem Fe located in the same place as in wild-type BFR. Nevertheless, the Met52His and Met52Leu variants of BFR assemble into stable 24mers.

The reduction potentials of the haem of *A. vinelandii* BFR at pH 8 is –475 mV in the presence of the non-haem iron core and –225 mV in its absence [25]. This shift in potential has been ascribed to a core electrostatic field effect as spectroscopic properties of the haem, which should indicate if there are substantial conformational changes accompanying core loading, are not significantly affected by the core [60]. Even in the absence of the core though, the reduction potential of the BFR haem is far lower than expected based on studies of bis-His-and His-Met-coordinated cytochromes [see 38 and references therein]. Coordination by Met sulphur gives rise to a reduction potential ~160 mV more positive than that provided by His N with all other factors being equal [64], a value similar to the ~180 mV increase in potential for the cytochrome b_{562} variants described by Barker et al. [63] in which the His-Met ligation was changed to bis-Met. Thus, with typical His-Met-coordinated cytochromes with reduction potentials in the range of 100–500 mV, the reduction potential of the bis-Met-coordinated haem of BFR is unusually low. This implies that factors other than inner-sphere coordination determine the level of the reduction potential of BFR, a situation similar to that of other metalloproteins [38, 65, 66 and references therein]. Though detailed calculations have not been reported for BFR it is most likely that the electrostatic environment of the haem strongly favours the Fe(III) state and that this factor far outweighs the Fe(II)-favouring effect of bis-methionine ligation.

4.2 Additional Haem Binding

The stoichiometry of haem binding to native wild-type BFR is variable with reports of 5–12 intrinsic haem groups bound per 24 subunits, depending on the source of the protein and the treatment it has undergone [1–3, 20, 24]. Also, for over-expressed recombinant BFR, haem levels as low as <1 per 24mer have been obtained. However, in some cases the addition of extra haem to a site that allows strong field coordination, as determined from the low spin character of the additional bound haem, was observed [67]. Initially this was considered to result from haem binding to native intra-subunit sites but further experimental investigations, coupled with the discovery of the inter-subunit binding site described above, led to a reevaluation of the in vitro binding studies which indicated that an ageing process was responsible for allowing the additional haem to be bound [60]. Additional haem could not be bound to freshly prepared recombinant BFR.

Similarly, though horse spleen ferritin has been shown to bind haem in vitro [68], at a bis-histidine coordination site which molecular modelling studies suggest are inter-subunit sites involving the 3-fold channels and histidines 114 and 124 (L-chain numbering) as ligands [69], and X-ray studies of horse spleen ferritin that had been treated with metalloporphyrins show that there is a porphyrin binding site in the protein shell [70, 71], at a 2-fold inter-subunit interface similar to that occupied by haem in BFR and betaine in crystals of amphibian red-cell L-chain ferritin [52], it appears that such binding is a result of an ageing process. A common feature of all these studies is that commercially available horse spleen ferritin, from Sigma, or aged recombinant ferritin was used. Ferritin freshly prepared from pony heart, and commercial horse spleen ferritin from other suppliers do not bind haem [J Knight, RJ Ward and GR Moore, unpublished data].

5 The Dinuclear Metal Site of Bacterioferritin

5.1 Is There a Dinuclear Metal Centre?

Each subunit of BFR contains a dinuclear metal-binding site (Figs. 3 and 5). This was first proposed by Cheesman et al. [53] based on similarities between BFR and H-chain ferritins, for which the proposal of intra-subunit dinuclear sites had some experimental support [51, 72–74], and confirmed by the crystallographic studies of Frolow et al. [21, 22] and the spectroscopic studies of Keech et al. [26, 75]. The latter studies have shown unequivocally that dinuclear metal centres of divalent ions can be formed in BFR but as yet there is no compelling evidence for the formation of dinuclear centres of trivalent ions. The possibility that a stable di-Fe(III) centre does not form in BFR has been taken by some to be significant in the function of the BFR dinuclear metal site, as will be discussed in see Section 8.

The central importance of the dinuclear centre illustrated in Fig. 5 to the function of BFR is brought into question by the report that the BFR from *Mycobacterium avium* has glutamine instead of glutamate as residue 18 [76]. However, it has not been established that *M. avium* BFR has a ferroxidase activity, nor what the activity is of *E. coli* BFR with a Glu18Gln mutation. The other residues illustrated in Fig. 5(a) are conserved amongst BFRs analysed to date.

In order to put the discussion of the BFR dinuclear metal centres into context we briefly describe the corresponding centres of ribonucleotide reductase (RNR) [77, 78] and methane monooxygenase (MMO) [79, 80], both of which have been crystallographically characterised. A schematic view of the di-Fe(II) and di-Fe(III) centres of RNR is shown in Fig. 5. There is obvious similarity between the dinuclear metal centre observed in crystals of BFR and the proposed di-Fe(II) centre of RNR that even extends to the approximate amino acid sequence separation between the coordinated residues. This similarity is all the more remarkable in that there is not a statistically significant overall amino acid sequence homology between BFR and RNR, so that it appears that their centres have evolved independently from each other. This similarity has been commented on previously [43, 77], and BFR assigned to the class II family of di-iron-oxo proteins [81]. The di-Fe(III) centre of RNR is substantially different from that of reduced RNR (Fig. 5), principally in the addition of a μ-oxo bridge between the two iron ions, which is coincident with a reduction in the inter-Fe separation, Asp84 becoming bidentate and Glu238 becoming monodentate, and the coordination of two water molecules, one to each Fe(III) ion [78]. Overall, the structures of the di-Fe(II) and di-Fe(III) centres of MMO markedly resemble those of RNR. Differences include the coordination of a bridging acetate ion in MMO in place of the bridging glutamate of RNR, though this could be a crystallisation artefact, and the presence of a μ-hydroxo bridge instead of a μ-oxo bridge [78–80, 82]. The 4-α-helical bundle structure of the BFR subunit (Fig. 3) is also common to the RNR and MMO domains that contain their di-

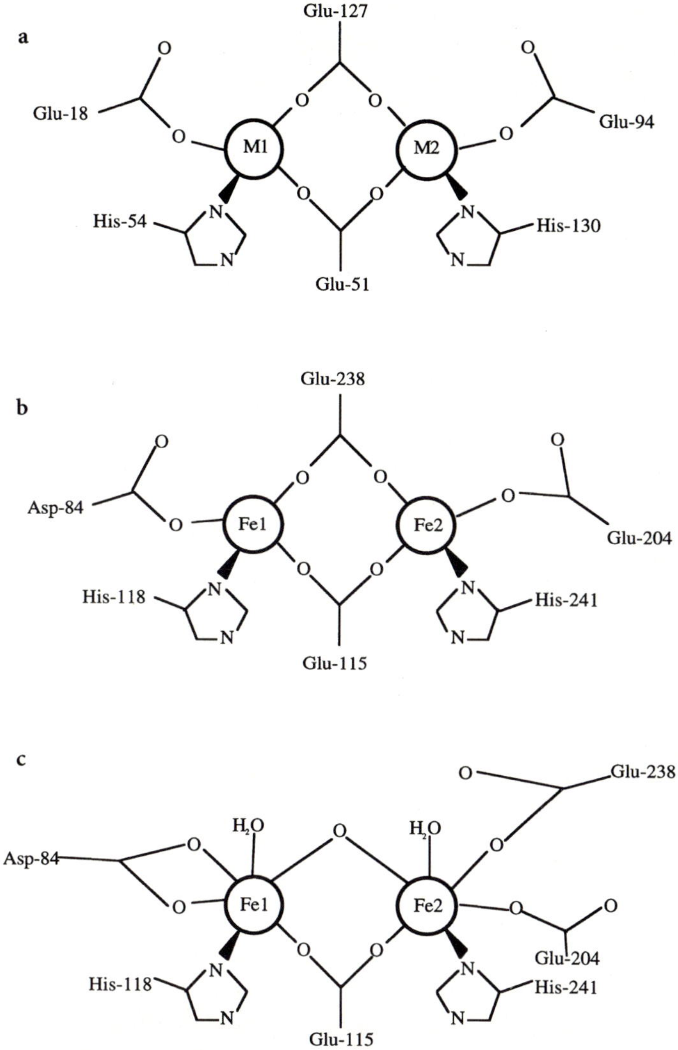

Fig. 5. Schematic representations of the dinuclear metal centres of BFR and the R2 protein of ribonucleotide reductase. **a** BFR. Although the chemical nature and redox states of the occupying metal ions were not stated by Frolow et al. [22] it is likely that M1 and M2 are both Mn(II), since crystals of BFR were obtained from solutions of apo-BFR and 100 μM $MnCl_2$. **b** Reduced RNR. **c** Oxidised RNR. (Constructed from information provided in references 22 and 78)

iron centres, and the disposition of the di-iron centres with respect to the long axes of the bundles is also similar [78]. Thus, structurally it is clear that RNR, MMO and BFR form a closely related family. However, given that it appears that a stable di-Fe(III) centre of the RNR/MMO type does not form within BFR (see Section 5.5) here similarity about the di-iron centre disappears. Understanding why this is so is a key problem within the di-iron protein field.

In order to analyse the different stages of Fe(II) uptake and oxidation (see Section 2(i)) investigation of the interaction of BFR with metal ions other than Fe(II) is helpful, as long as these metal ions have different redox stabilities and binding affinities to Fe(II), or are spectroscopically more easily detectable. Mn(II), Co(II) and Zn(II) fall into these categories and thus the following sections describe their use to probe the reactivity of the ferroxidase centre.

5.2
Binding of Mn(II)

The X-ray structure of BFR shows one centre in each subunit, containing a pair of metal ions (Fig. 3). The structure of this centre strongly resembles the di-Fe(II) centre of RNR (Fig. 5) and its di-Mn(II) homologue which can be prepared from apo-RNR [83]. The structure of the BFR centre is rather symmetrical, each metal ion having the same number and type of ligation. The symmetrical nature of the centre suggests that the two metal ions are divalent, though the original X-ray work did not identify them [22]. Since the BFR was crystallised in the presence of 100 μM $MnCl_2$ it seems likely that the site is occupied by a pair of Mn(II) ions.

Potentiometric studies with apo-BFR, in which the number of protons released per metal ion added was determined [26], show clearly that 48 Mn(II) ions bind to each 24mer with the release of ~2 protons per metal ion, and apparent dissociation constants for each Mn(II) ion of $\sim 10^{-4}$ M. EPR titrations confirm the binding stoichiometry [26]. Since similar studies of Co(II) interactions with BFR have shown that the dinuclear centre is occupied before any other site, with 1.6–2.0 protons released per divalent metal ion [84], the potentiometric binding data for Mn(II) suggest that Mn(II) binds at the same dinuclear site as Co(II) but with a lesser affinity.

5.3
Binding of Co(II)

Optical and EPR spectroscopies reveal that Co(II) binds to the dinuclear site of BFR [75]. This conclusion follows from EPR studies of Co(II)-substituted wild-type and haem-free BFRs that indicate two cobalt ions are weakly magnetically coupled resulting in a broad EPR signal with pronounced ^{59}Co hyperfine splitting (Fig. 6), as well as from optical studies of the d-d absorption bands of Co(II) bound to wild-type, haem-free (Fig. 7) and other mutant forms of BFR which lack key binding residues at the dinuclear site (Table 2). Studies of the dinuclear centre variants Glu18Ala and Glu94Ala BFR also suggest that the metal-binding stoichiometry is controlled by the charge at the binding site, in addition to providing confirmation of the site of Co(II) binding. In both variants the replacement of one potentially negatively charged residue at the binding pocket by a neutral residue reduced the binding stoichiometry to one divalent metal ion per BFR subunit, with a dissociation constant similar to that of the wild-type binding pocket. However, it is necessary to replace the glutamates by histidine to be certain that it is the charge at the site and not the number of ligands that determine stoichiometry.

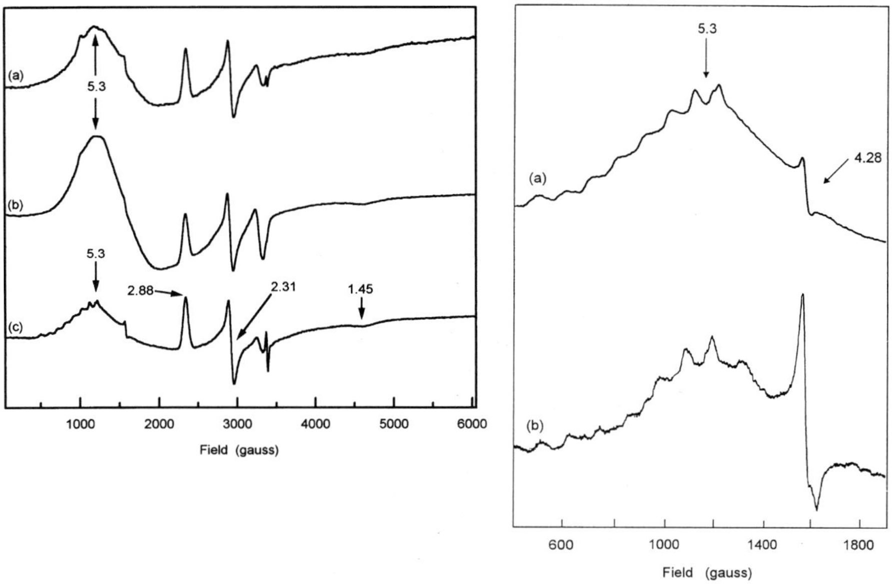

Fig. 6. Low-temperature X-band EPR spectra of BFRs containing Co(II) at the intrasubunit dinuclear sites. *Left*: **a** E18A BFR (8.6 μM) containing 24 Co(II) ions per 24mer; **b** E28A BFR (8.6 μM) containing 48 Co(II) ions per 24mer; **c** Wild type BFR (18.6 μM) containing 48 Co(II) ions per 24mer. [Buffer was 100 mM MES, pH 6.5. Measurement conditions: Microwave frequency 9.4 GHz; Modulation 100 KHz; Modulation amplitude 10 Gauss; Temperature 5 K; Microwave power 2.01 mW]. *Right*: Low field regions of the spectra of Co(II)-substituted **a** wild-type and **b** M52L BFRs, both containing 48 Co(II) ions per 24mer. [The M52L BFR was 21.7 μM 100 mM HEPES buffer, pH 7.1] (Reproduced from reference 75)

Table 2. Absorption characteristics, dissociation constants and binding stoichiometries for Co(II) added to *E. coli* wild-type and variant bacterioferritins. [Data from reference 75]

Bacterioferritin	K_d ($\times 10^5\ M^{-1}$)	Number of Co(II) bound per BFR	Absorption characteristics λ_{max} nm ($\varepsilon\ M^{-1}\ cm^{-1}$)
Wild type BFR			
with $CoCl_2$	1.4 ± 0.3	45.0 ± 0.6	520 (126), 555 (155), 600 (107), 625 (75)
with $Co(ClO_4)_2$	1.1 ± 0.2	45.0 ± 0.6	520 (133), 555 (162), 600 (111), 625 (75)
M52H BFR	0.96 ± 0.5	48.4 ± 0.3	520 (120), 555 (155), 600 (111), 625 (74.4)
E18A BFR	1.05 ± 0.2	29.0 ± 0.9	510 (155), 550 (165), 590 (200)
E94A BFR	0.88 ± 0.2	19.3 ± 3	510 (123), 552 (171), 595 (138), 620 (116)

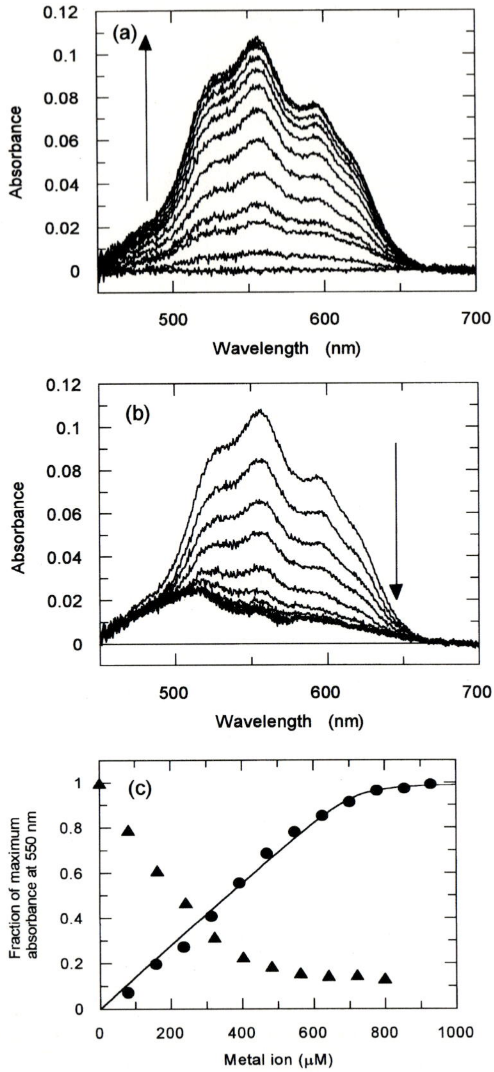

Fig. 7. Optical titrations of apo-BFR with solutions of Co(II) and Zn(II) chlorides. Apo-BFR (13.2 μM as 24mer) was in 0.1 M MES buffer, pH 6.5. Visible difference spectra were recorded with apo-BFR solutions in both reference and sample cuvettes, and with metal-containing solutions added to the sample cuvette and an equal volume of water added to the reference cuvette. **a** Difference spectra recorded after the addition of each aliquot of Co(II) to apo-BFR. **b** Difference spectra recorded after the addition of each aliquot of Zn(II) to an apo-BFR solution containing Co(II) (930 μM). **c** Change in absorbance with increasing concentrations of Co(II) (circles; 0–930 μM) or Zn(II) (triangles; 930 μM Co(II), 0–790 μM). The shape of the Zn(II) curve indicates that Zn(II) displaces Co(II) at low concentrations of Zn(II) even though Co(II) is present in a large excess over the number of dinuclear centres. (Reproduced from reference 84).

The substitution of Co(II) for Mn(II) is not expected to greatly affect the structure of the centre and thus Fig. 5 indicates that Co(II) ions bound at the two ends of the site should have very similar optical properties; something the experimental data agree with, though they do not distinguish between distorted tetrahedral and pentacoordinate geometries. The relatively large extinction coefficients (Table 2) rule out octahedral coordination, however [75 and references therein].

The binding of Co(II) was found to be positively cooperative but conditions under which the optical titrations were performed were found to be crucial for the detection of this cooperativity. Under conditions in which the average K_d was much less than the concentration of Co(II)-binding sites, cooperativity was largely obscured but it became more readily detectable at lower protein concentration.

5.4 Binding of Zn(II)

Direct evidence for Zn(II) binding to the intra-subunit dinuclear metal sites of BFR is lacking but indirect evidence is convincing. Thus, Zn(II) displaces Co(II) bound at the dinuclear sites [84] and competes effectively for rapid Fe(II) oxidation sites on BFR [43], which mutagenesis studies indicate are the dinuclear sites. Potentiometric experiments confirm that two Zn(II) ions can bind per subunit, with an apparent K_d of 1×10^{-7} M [84].

5.5 Binding of Fe(II) and Fe(III)

Studies with the ferrous form of the ferroxidase centre are hampered by the lack of easily detectable EPR and optical signals from Fe(II), and by the reactivity of the centre towards oxygen. Hence, direct spectroscopic evidence for the dinuclear iron centre is elusive. The fact that intra-subunit dinuclear centres are formed from Mn(II) and Co(II) has led to the assumption that similar Fe(II) centres are formed.

Evidence for a stable di-Fe(III) centre being formed after oxidation of Fe(II) is lacking. EPR measurements indicate that shortly after oxidation of the di-Fe(II) centre mononuclear Fe(III) is present in relatively high amounts [44]. This suggests that the di-Fe(III) centre is unstable, which is a major difference between BFR and other dinuclear iron proteins. In other proteins, such as MMO and RNR (Fig. 5), there are μ-oxo or μ-hydroxy bridges between the two Fe(III) ions which lead to diagnostic magnetic coupling of varying strength. The EPR data for BFR indicate that the Fe(III) ions are not magnetically coupled, and thus they rule out a μ-oxo or μ-hydroxy bridged di-Fe(III) centre, though they do not eliminate two non-coupled Fe(III) ions bound to the site illustrated in Fig. 5. Whichever explanation accounts for the EPR observations, there is a major difference between BFR and other dinuclear iron proteins. Other techniques, such as ^{57}Fe Mössbauer spectroscopy or X-ray crystallography, are needed to further define the Fe(III) centres resulting from oxidation of the di-Fe(II) complex of BFR.

5.6 Electrostatic Aspects of Metal Ion Binding

A key observation in studies of the binding of divalent metal ions to apo-BFR at pH6.0–6.5 is that there is partial or complete charge compensation to the binding interactions [26, 84]. Thus binding of Mn(II), Co(II) and Zn(II) ions is accompanied by the release of ~2, ~1.6 and ~2 protons per metal ion bound, respectively. Coupled with the finding that the Glu18Ala and Glu94Ala variants of BFR only bind one divalent metal ion at each of their ferroxidase centres [75], this suggests that electrostatic factors at the dinuclear centre are important for divalent metal ion binding and oxidation.

The apparent instability of the dinuclear iron centre as a bridged dimeric Fe(III) centre might also be a result of electrostatic factors, at least in part. In proteins such as RNR and MMO, the dinuclear Fe(III) centres are stabilised by the formation of μ-oxo or μ-hydroxo bridges between the two Fe(III) ions. These, together with the charge states of the amino acid ligands to the Fe(III) ions, give partial or complete charge compensation to the six positive charges. However, if formation of a μ-oxo bridge is avoided in BFR then charge compensation at the ferroxidase centre cannot be achieved for two Fe(III) ions. Such a difficulty could be translated into structure-function relationships in a number of ways, including the di-Fe(III) centre being electrostatically unstable so that it breaks down almost immediately after it is formed, possibly expelling Fe(III) into the central cavity to initiate core growth.

5.7 Di-Iron Centres in Human H-Chain Ferritin and *E. coli* FTN

We note that *E. coli* FTN [19] and human H-chain ferritin subunits [8] also contain dinuclear metal binding sites, though there are significant differences compared to the BFR site (Fig. 5). In both FTN and H-chain ferritin one of the two histidine residues is replaced by glutamic acid, and one of the bridging glutamates is replaced by glutamine. Thus it appears that the metal-coordinating side-chains at these sites could give a total site charge of –4, the same as at the corresponding BFR sites [84]. Somewhat surprisingly then, the X-ray structure of FTN, elucidated with crystals obtained under conditions where the two ions bound at the dinuclear centre should both be Fe(III) ions, failed to identify either a μ-oxo or a μ-hydroxo bridge [19]. This might be a result of the iron ions bound at the dinuclear site having been reduced in situ by the X-ray irradiation (Hempstead & Harrison, personal communication), as described by Logan et al. [78] for RNR. Another possibility is that long-range electrostatic interactions help stabilise the di-Fe(III) centre of FTN so that a μ-oxo or μ-hydroxo bridge is not necessary to stabilise it. However, there is no precedent for this. The di-Fe(III) site of H-chain ferritin has not been characterised by crystallographic methods, but ^{57}Fe Mössbauer spectroscopy suggests it contains either a μ-oxo or a μ-hydroxo bridge [55, 73, 74]. Similar Mössbauer studies of FTN have not been reported.

5.8 Interaction Between the Haem and Dinuclear Centres

There are clear spectroscopic data which show that the properties of the haem groups and dinuclear centres are interlinked [43, 44]. This interaction could be due partly or solely to electrostatic influences of the bound metal ions, or a result of conformational changes accompanying the metal ion binding. However, whatever the link, the observation that aerobic iron-uptake rates into BFR are independent of the presence or absence of haem indicates it is not a major functional feature for this reaction [58]. That these centres are linked is perhaps not too surprising though, since they appear to form discrete structural units with sequential residues acting as ligands to the different centres, Glu51 to the dinuclear centre and Met52 to the haem group (Fig. 8).

The first indication of the linkage came from the finding that binding of Fe(II) to apo-BFR, phase I of the aerobic oxidative uptake of Fe(II) (see Section 2), caused a ~1 nm shift of the haem Soret absorption to higher energy [44]. Thus Fe(II) binding at the dinuclear centre influences the haem. Subsequently it has been shown that a similar haem spectral perturbation results from the binding of other divalent ions to the dinuclear centre [43, 75]. Surprisingly then, detailed analysis of the optical titration data for Co(II) binding to apo-BFR indicates that its K_d is largely independent of the presence or absence of the haem groups [75]. This result is even more surprising when the displacement by Zn(II) of Co(II) from apo-BFR, in which only 40–50% of the haem binding sites were occupied, is considered [84]. Even at saturating

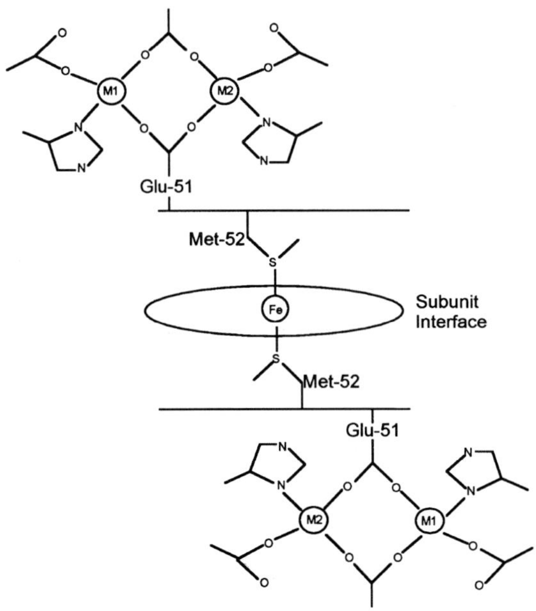

Fig. 8. Schematic representation of the structural relationship of the dinuclear centres and the haem groups of BFR

levels of Zn(II) considerably in excess of the total Co(II) concentration, less than 100% of the Co(II) bound at the dinuclear site is displaced by Zn(II): visible absorption difference spectra in the presence of saturating levels of Zn(II) showed an absorption that can be ascribed to a Co(II)-BFR complex (Fig. 7). This absorption does not result from displaced Co(II) ions present as $[Co(H_2O)_6]^{2+}$, and since the form of the spectrum is different to that of the Co(II)-BFR complex in the absence of Zn(II) it most probably arises from Co(II) bound to some of the dinuclear centres in which Zn(II) is also bound. This implies that the dinuclear centres do not all behave the same in this experiment. It may be that centres with an adjacent haem are able to bind two Co(II) ions, but only one Zn(II) ion in the presence of competing Co(II). The relevant binding affinities, Co(II) > Zn(II), then run counter to the Irving-Williams order. This aspect of metal binding to apo-BFR requires further investigation, particularly in view of the order of binding affinities to the dinuclear centres of apo-BFR determined potentiometrically: Mn(II) < Co(II) < Zn(II). This is the Irving-Williams order [85], a relationship frequently observed for metalloproteins; e.g. for the binding of Co(II) and Mn(II) to the dinuclear metal binding site of the 4-α-helical bundle apohemerythrin [86].

6 Mononuclear Sites of Bacterioferritin

Low-affinity binding sites for ~48 Zn(II) and ~48 Co(II) to the 24mer BFR have been detected potentiometrically [84], though there is no supporting spectroscopic evidence for this. Le Brun et al. suggested that this binding was most probably associated with adventitious binding to the surface of the negatively charged BFR, since the bound Co(II) had no observable d-d absorption, indicating an octahedral coordination, and an EPR spectrum indistinguishable from that of free Co(II). However, low-affinity binding of Mn(II) could not be detected, either by EPR spectroscopy or potentiometrically [26], and this is consistent with the X-ray structure of apo-BFR in the presence of 100 μM Mn(II) which revealed Mn(II) binding only at the dinuclear site [22]. Presumably, if the Mn(II) binding affinity for additional sites follows a similar trend to that for the dinuclear site, then their K_d values will be ~10^{-2} M, which is too high for significant binding to occur under the experimental conditions reported.

As with the dinuclear site, the binding of Fe(II) and Fe(III) to mononuclear sites has not been well explored due to experimental difficulties. EPR investigations have suggested that mononuclear Fe(III) is present almost immediately after Fe(II) has been oxidised at the dinuclear centre but the exact location of this Fe(III) has not been established [44].

7 Crystalline and Amorphous Polynuclear Iron Clusters

7.1 Composition

The central cavity of BFR houses polynuclear iron clusters that constitute the non-haem-iron core. It has a sufficient volume to accommodate 4500 iron atoms as ferrihydrite, $Fe_2O_3 \cdot H_2O$, [5] but as isolated BFR usually contains only 800–1500 atoms of iron per molecule (Table 1). In contrast, animal ferritins generally contain 1000–3000 atoms of iron per molecule in the isolated form, though the mean core sizes determined by electron microscopy are not very different from those of BFR [3, 28, 87]. Thus the two families of natural core have very different densities and compositions. When isolated the cores of all ferritin-type proteins are generally polydisperse with respect to their size. This can complicate studies of their composition and properties and therefore needs to be borne in mind in the following discussion.

7.2 Structure

The generally accepted structure for the crystalline iron core of ferritin is the ferrihydrite structure proposed by Towe and Bradley [88, 89]. This structure consists of oxygen layers with iron in octahedral sites between the layers. Powell [90] summarises the evidence in support of this model for the structure of the ferritin core, but it is important to note that the Towe and Bradley model is not accepted by all researchers for the mineral material [91 and references therein]. The BFR core is generally termed an "iron-oxy-hydroxide-phosphate", but it seems likely that different BFRs have somewhat different cores. This follows from the observation that the core of *A. vinelandii* BFR has a Mössbauer blocking temperature of ~20 K while that for *P. aeruginosa* BFR fails to show behaviour for which a blocking temperature can be recorded (see Sections 7.3 and 7.4).

Usually the difference between BFR-type cores and their animal ferritin counterparts is ascribed to the variation in phosphate content, with ferritins having a lower phosphate content than BFRs (Table 1). In vitro, cores with a higher degree of crystallinity and a lower phosphate content can be incorporated into BFRs. This is consistent with an important role for phosphate and shows that the core morphology is not determined solely by the protein shell [87, 92]. However, work on chiton ferritin, such as that from *P. laticostata*, which has a low phosphate content and an amorphous core [11], indicates that there is not a clear correlation between phosphate content and morphology. Nevertheless, it appears that the composition of the medium within which the cores are laid down influences the nature of the core.

The structures of the "iron-oxy-hydroxide-phosphate" cores of both native BFRs and ferritins, and those with cores reconstituted in vitro, have been examined using EXAFS methods [92, 93]. Studies of native *A. vinelandii* BFR have shown that there are fewer Fe-Fe contacts than with native horse ferritin, and those that do occur are at a greater interatomic distance [92]. This is consistent with EXAFS data from the same study showing that the core iron of BFR had 5–6 phosphorus atoms no more than 3.17 Å apart, thus supporting a model for the core of an amorphous Fe(III)-phosphate complex in which some of the phosphate bridges Fe(III) ions and some is non-bridging.

7.3 Magnetism and Spectroscopy

The core particles possess exchange-coupled lattices in which the electron spins of the individual high-spin Fe(III) ions are anti-ferromagnetically coupled. The spins become disordered above a certain temperature, T_{ord}, which is determined by the strength of the exchange coupling between the spins. For lattices with oxide bridges between Fe(III) ions the ordering temperatures are relatively high. Spins which have no partner with which to pair, termed uncompensated spins, occur at the surface of the particle. In a single domain particle they give rise to a net paramagnetic moment. In the case of a particle with uniaxial symmetry, the magnetisation vector of length S can take up a number of orientations relative to this unique axis, often called the easy axis, such that the projection of the spin along the axis varies from $-S$, $(-S+1)\cdots$ to $+S$. The anisotropy of the core particle separates the energy levels of the spin sub-levels which, in the case of a uniaxial field, leads to a ladder of doubly degenerate energy levels $\pm S, \pm(S-1)$ and so on. However, this creates a barrier to thermal transitions between levels with $+S, +(S-1)\cdots$ and $-S, -(S-1)\cdots$. The energies of the different levels of the spin manifold are given by:

$$E = \pm D(M_S)^2 \tag{5}$$

where D is the axial zero-field parameter which measures the extent of the anisotropy and M_s runs from $-S$ to $+S$. Hence the barrier height is DS^2. If the sign of the anisotropy constant is negative the resulting set of energy levels are as shown in Fig. 9. For temperatures at which kT is greater than the barrier height the spin vector can flop from $-S$ to $+S$. The core then exhibits thermal paramagnetism. At low temperature the spins become frozen into the minima and cannot flop from one orientation to another, the system is said to be blocked. The blocking temperature, T_B, is related to the barrier height and depends, according to equation (5), on the axial zero field splitting parameter, often called the anisotropy constant, and on the total number of uncompensated spins. This, in turn, is dependent on the size, as determined by the particle volume, and the crystallinity of the core particle, as well as its chemical nature. Thus the type of iron species present, whether ferrihydrite, goethite, magnetite or "iron-oxy-hydroxide-phosphate", is also important. It can sometimes be determined by electron diffraction. However, it can been seen that if there is a spread of particle sizes in a sample of ferritin there will be a distribution of spin magnitudes and hence of barrier heights and blocking temperatures.

Relaxation from one spin manifold to the other can be very slow (of the order of months) at temperatures well below the barrier height. High-order multi-

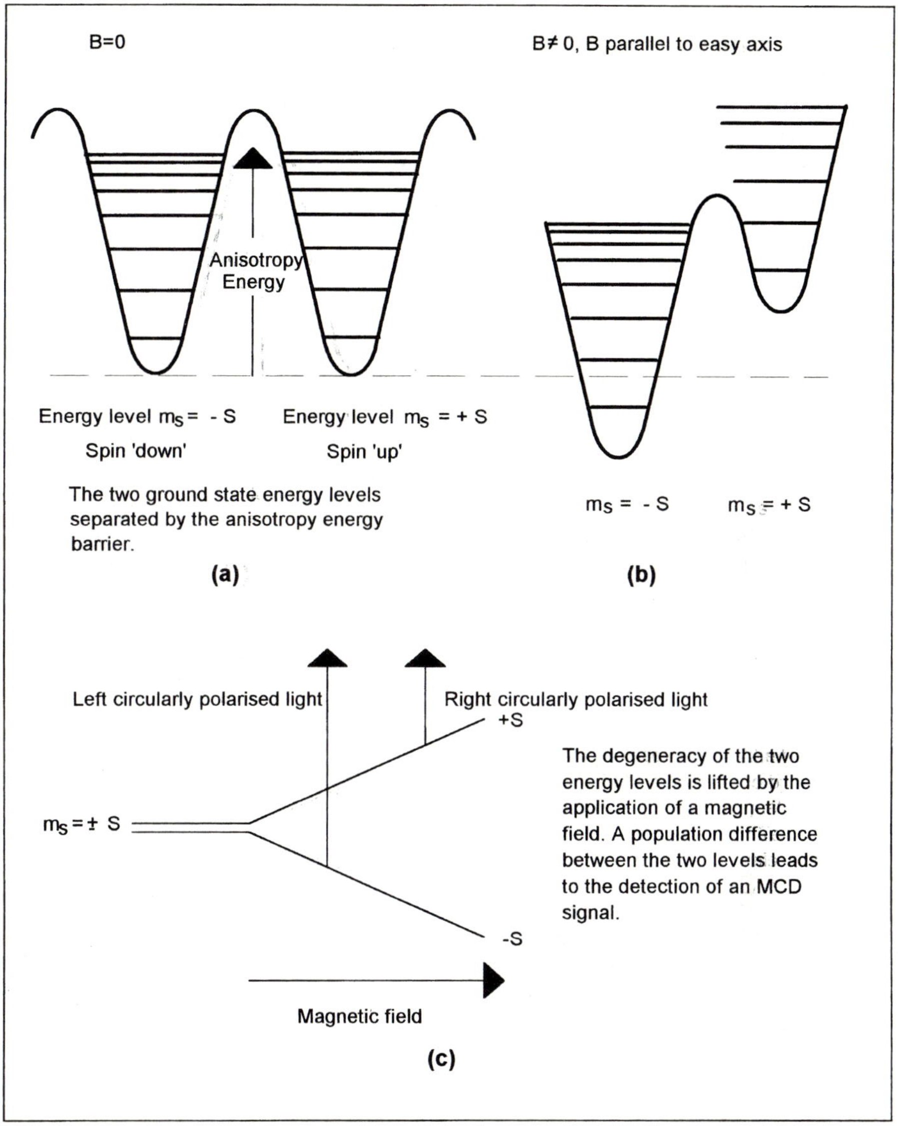

Fig. 9. Superparamagnetism. Schematic illustration of the ground state energy levels of uncompensated electron spins in ferritin cores **a** in the absence and **b** in the presence of a magnetic field applied parallel to the easy axis. The total spin vector is S and the energy levels are separated by $[D(S-n)^2-D(S-n-1)^2]$ where n runs from 0 to S. The height of the barrier is DS^2 where D is an axial zero-field splitting related to the axial anisotropy (see text). **c** The MCD signal arises because transitions from pairs of level $\pm$ S, $\pm$ (S−1), etc are oppositely circularly polarised. Hence a preponderence of population in one of these pairs will lead to a net absorption of circularly polarised light

poles arising from anisotropy in the plane perpendicular to the unique axis are required to induce transitions between the +S and –S manifolds. However, evidence has recently been presented for quantum mechanical tunnelling across the barrier as a significant contribution to cross-relaxation in the molecular magnet $[Mn_{12}O_{12}(OAc)_{16}(H_2O)_4]$, which has a total spin S = 10 and an axial zero-field splitting of -0.6 K. This gives a barrier height of 60 K [94–96]. These publications also consider the possibility of quantum tunnelling in ferritin cores, though it is clear that additional experimental data are required to properly address this question.

Application of a magnetic field has two effects. First the energies of the two spin subsets are changed, those with +S components being raised in energy and those with –S lowered. Therefore spin relaxation from one subset of spins to the other can occur. The presence of a thermal barrier to the reorientation of the spins also leads to a hysteresis in the magnetic susceptibility or magnetisation when the magnetic field is swept to zero from the value needed to obtain full magnetisation. A magnetic field applied perpendicular to the unique axis may also cause spins to relax across the barrier.

Blocking temperatures have been measured in both ferritins and BFR using ^{57}Fe Mössbauer spectroscopy [11, 25, 27, 97, 98] (see Section 7.4). More recently [99] it has been shown that variable-temperature MCD enables the hysteresis of ferritin cores to be observed since residual magnetisation remains in a sample below the blocking temperature after application of a saturating magnetic field and hence circular dichroism is apparent in a zero magnetic field (see Section 7.6).

The magnetic properties of the naturally occurring amorphous cores of BFR have not been as thoroughly explored as have those of the crystalline ferritin cores. Thus, while magnetic susceptibility measurements of ferritin show that it has a residual magnetic moment at room temperature of ~ 3.8 BM for each iron resulting from the uncompensated spins of the high-spin Fe(III) ions [100], comparable data for bacterioferritin have not been reported. A major difference in the magnetic properties of crystalline cores of the type found in native ferritins and the amorphous cores of the type found in *P.aeruginosa* BFR is that the former exhibit superparamagnetic behaviour with blocking temperatures of 30 to 40 K while the latter do not (Table 1). Below we consider the experimental evidence that supports this assertion.

7.4
^{57}Fe Mössbauer Spectroscopy

^{57}Fe Mössbauer spectroscopy has been widely employed to study ferritins and BFR [e.g. 11, 25, 27, 34, 101, 102]. The spectra of the ferric cores are temperature-dependent and consist of a quadrupole split doublet at temperatures above the blocking temperature and a magnetically split sextet at temperatures below this (Fig. 10). This is in agreement with the picture given above. Above the blocking temperature the spins relax from one orientation to another at a fast rate compared with the Mössbauer frequency and hence the Mössbauer nucleus does not experience the internal field anisotropy. Below the blocking temperature

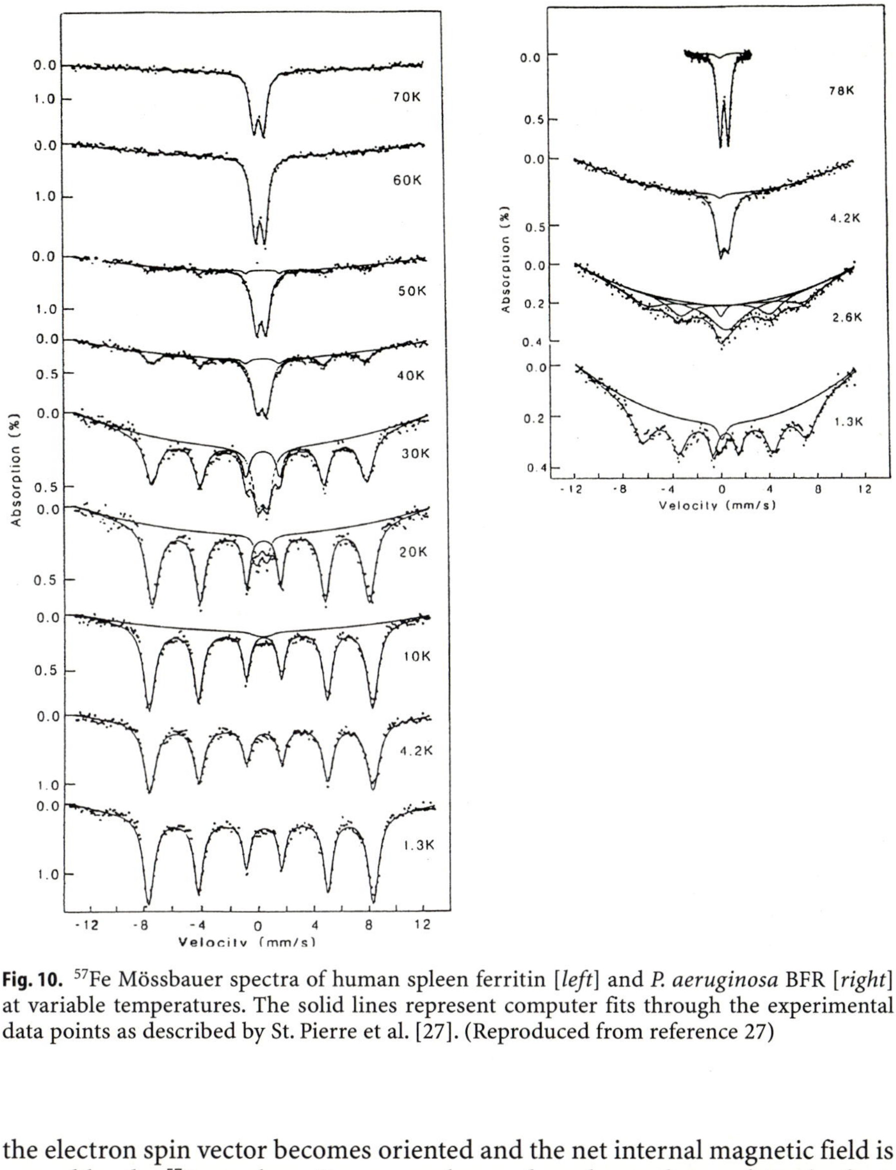

Fig. 10. ^{57}Fe Mössbauer spectra of human spleen ferritin [*left*] and *P. aeruginosa* BFR [*right*] at variable temperatures. The solid lines represent computer fits through the experimental data points as described by St. Pierre et al. [27]. (Reproduced from reference 27)

the electron spin vector becomes oriented and the net internal magnetic field is sensed by the ^{57}Fe nucleus. Two general cases have been observed. In the first, the magnetic ordering temperature (T_{ord}) is greater than the blocking temperature (T_B), and in the second $T_B > T_{ord}$. Proteins in the former class, exemplified by horse ferritin with a T_B of ~40 K, exhibit superparamagnetic behaviour. The latter class, illustrated by *Pseudomonas aeruginosa* bacterioferritin with a $T_B > 3$ K and a $T_{ord} < 3$ K, does not show superparamagnetic behaviour. Thus ^{57}Fe Mössbauer spectra are consistent with electron microscopy data indicating that the cores of bacterioferritin are largely amorphous whilst those of animal ferritins are largely crystalline.

7.5 EPR Spectroscopy

EPR spectra reveal differences between the ferritin cores and bacterioferritin cores which contain phosphate. At 4 K bacterioferritin cores are EPR-silent [61], but as the temperature is raised a broad isotropic signal at g ~ 2 appears which becomes narrower as the temperature increases (Fig. 11 [left]). The origin of this signal is not certain but since BFR cores with 800 iron atoms per 24mer and 24–68 iron atoms per 24mer have a similar EPR signal, which is absent from core-free protein, it must arise from the core particle [61, 103]. The signal appears to be reporting the relaxation characteristics of the core and may arise either from uncompensated spins or, more likely, from uncoupled spins since the exchange coupling temperature is below 4 K. Ferritin cores give signals at high temperatures at g ~ 6 and g ~ 2.2 (features A and B in Fig. 11 [right]) that are lost at 4 K [104, 105]. It has been assumed that these EPR signals originate from the uncompensated electron spins of the antiferromagnetically ordered cores and if this is so there should be a correlation between the EPR and

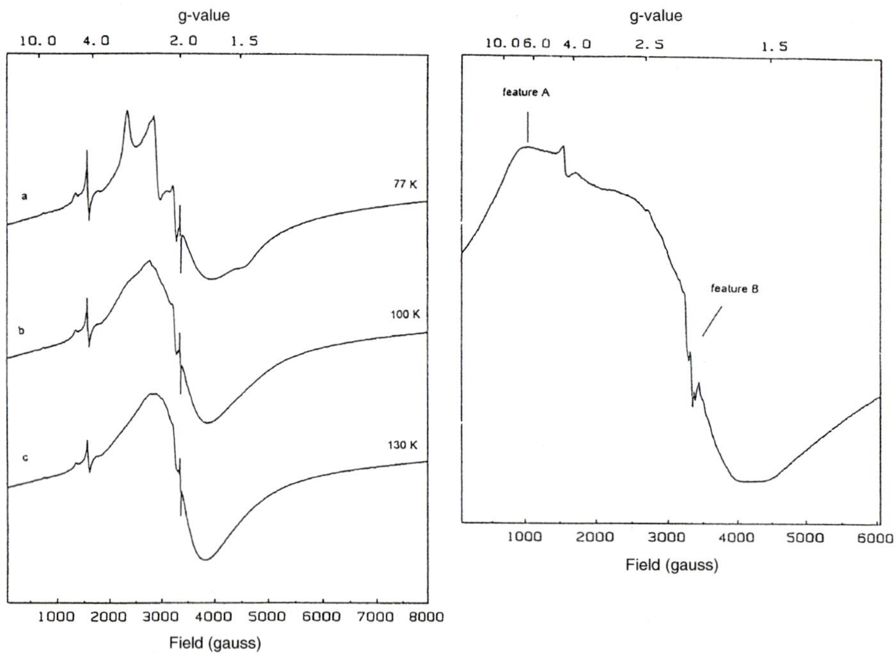

Fig. 11. X-band EPR spectra of *E. coli* BFR [*left*] and horse spleen ferritin [*right*]. The BFR was 120 μM as a 24mer, contained an iron core of ~50 Fe(III) ions per 24mer and was in a 1:1 mixture of 0.1 M sodium phosphate buffer, pH7, and glycerol. The ferritin was 100 μM as a 24mer, contained an iron core of ~1100 Fe(III) ions per 24mer and was dissolved in 0.1 M sodium phosphate buffer, pH7. (Reproduced from reference 103) (Note that both sets of spectra contain signals at g ~ 4.3 from monomeric Fe(III) and sharp signals at g ~ 2 that may arise from free radical species. The BFR spectrum at 77 K also contains signals from the ferrihaem at g = 2.88 and g = 2.31 that are too broad to be clearly resolved at higher temperatures)

Mössbauer data. This appears to be the case for animal ferritin, which Mössbauer indicates is magnetically ordered at 4 K.

7.6 Magnetic Circular Dichroism Spectroscopy

MCD spectra of amorphous BFR and crystalline ferritin cores in magnetic fields up to 5 T and at a temperature of 1.6 K have been reported [99]. Thus BFR core transitions gave MCD spectra typical of other paramagnetic Fe(III) proteins, no features characteristic of superparamagnetic behaviour were observed. In contrast ferritin cores gave spectra that showed a hysteresis of the MCD intensity during magnetic field cycling at 1.6 K which was absent at 77 K, above the blocking temperature (Fig. 12).

This hysteresis is due to the superparamagnetism of the core. At 1.7 K spin relaxation between the spin manifolds with +S and –S is very slow and so a net population is generated in one spin sub-level after magnetisation by a strong field. Optical transitions out of this level are circularly polarised. Hence an MCD signal is seen. On reversing the polarising magnetic field, the population is switched to the opposite subset of spin sub-levels and so the circular polarisation of the optical transition is reversed. This provides a sensitive optical method for monitoring the characteristic parameters of the core magnetism.

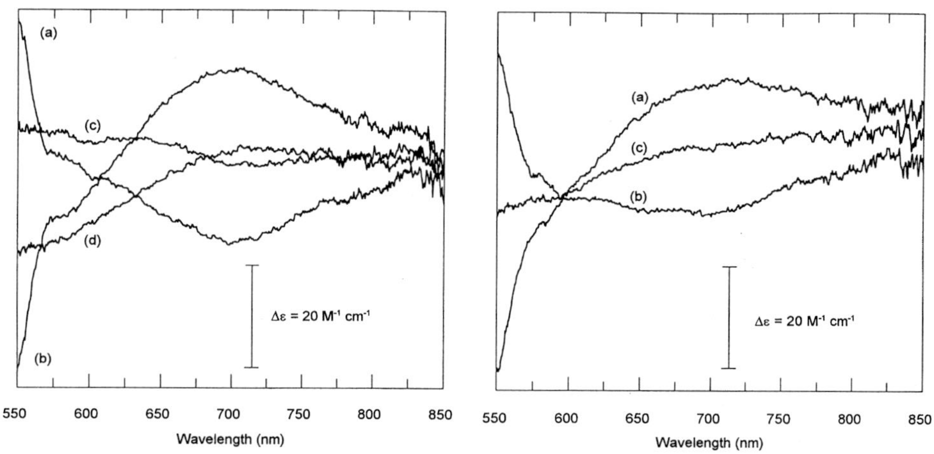

Fig. 12. MCD spectra of horse spleen ferritin. The sample was 100 μM ferritin 24mer containing, on average, 1100 Fe(III) ions per 24mer in a 1:1 mixture of 0.1 M Na phosphate buffer, pH7, and glycerol. *Left panel:* spectra recorded at 1.6 K with **a** a forward magnetic field of 5T; **b** a reverse magnetic field of 5T; **c** zero magnetic field following the application of a reverse magnetic field; and **d** zero magnetic field following the application of a forward magnetic field. *Right panel:* spectra recorded at 77 K with **a** a forward magnetic field of 5T; **b** a reverse magnetic field of 5T; and **c** zero magnetic field. The key observation is that at 1.6 K the spectrum in zero magnetic field is dependent on the direction of the external magnetic field applied previously, and at 77 K the spectrum in zero magnetic field is independent of the magnetic history of the sample. (Reproduced from reference 99)

The method has an additional selectivity since the polarisation of the optical transitions is important in determining which sub-population of ferritin molecules can be detected. This remains to be exploited in studies of core magnetism.

Assignment of the core UV-visible bands observed in the MCD spectra of ferritin and BFR has not been reported. Fe(III) ions octahedrally coordinated by O^{2-} ions generally have spin-forbidden d-d absorption bands in the range 500–1000 nm that arise from transitions such as those from the $^6A_1$ to ^4G free ion states $^4T_1$, $^4T_2$, ^4E and ^4A, or the ^4D free ion states $^4T_2$ and ^4E. Exchange interactions in magnetically coupled lattices may break the spin-forbidden selection rule of such transitions, thereby increasing absorbance intensities and hindering a straightforward assignment of the bands. As a result of this Le Brun et al. [99] did not attempt to assign the MCD bands although they did note that Webb and Gray [106] had assigned a band at 900 nm in the absorption spectrum of ferritin with a crystalline core to the $^6A_1 \rightarrow {}^6T_1$ (^4G) transition of octahedral Fe(III).

7.7 Redox Potentials

The reduction potential of the core iron of *A. vinelandii* BFR has been determined electrochemically to be -420 ± 20 mV at pH 7–9 [25]. This is markedly lower than the reduction potential of the core iron of horse spleen ferritin, which has been shown to be -190 mV at pH 7, a value which decreases with increasing pH [50]. Presumably the difference in pH dependence reflects the much higher loading of phosphate in the core of BFR compared to ferritin (Table 1), with a consequent reduction in the oxide content of the core, but the difference in the actual levels of the potentials could reflect a variety of factors, both short-range within the core itself and longer-range influences of the protein shell.

8 Mechanistic Roles of the Metal Centres

8.1 Iron Uptake

Aerobic oxidative uptake of iron by BFR involves at least three kinetically distinguishable phases (Fig. 1) corresponding to: binding of Fe(II) at the ferroxidase centre (phase 1); fast oxidation of Fe(II) to Fe(III) at the ferroxidase centre (phase 2); and subsequent formation of the mineral core (phase 3) [43, 44] (see Section 2.1). The haem groups do not appear to play a major role in this process as they do not cycle their oxidation state during oxidative iron-uptake, and haem-free variants of BFR have similar iron-uptake rates to those of the wild-type protein [58]. However, the optical spectrum of the haem responds to the binding of metal ions at the ferroxidase centre and some of the binding properties of divalent metal ions at the ferroxidase centre appear to be influenced by the haem (see Section 6). Nevertheless, we shall not consider the haem further in this discussion of iron uptake but focus on the dinuclear metal centre in each subunit.

The presence of cooperativity in divalent metal-binding to the ferroxidase centre of BFR (see Section 5) is likely to be of mechanistic significance. It implies that upon addition of Fe(II) to apo-BFR the ferroxidase centre preferentially takes up a pair of divalent metal ions. Hence oxidation takes place at a filled ferroxidase centre and the reduction products of dioxygen are then likely to be even-electron species, such as hydrogen peroxide or water, rather than radicals such as superoxide, as shown for animal ferritin (see equations 1 and 2) [45, 46].

A variety of roles can be envisaged for the dinuclear centre, some of which are illustrated in Fig. 13 [107]. If the dinuclear centre were stable in both its oxidised and reduced states, as the corresponding centres of MMO and RNR are, then a *Linked Transfer* mechanism could operate in which the first stage of iron uptake into apo-BFR is the formation of the dinuclear centre at which O_2 reduction will occur (see Fig. 13(a)). However, a *Sequential Transfer* mechanism could be operative in which all the iron that makes up the core must pass through the ferroxidase centre (Fig. 13(b)). For the *Linked Transfer* mechanism, the dinuclear centre would be stable but there is no evidence that a stable di-Fe(III) centre is formed in BFR (see Section 5). However, the *Sequential Transfer* mechanism portrayed in Fig. 13(b) also fails to adequately account for the experimental data. Phase 2 of the kinetic profile of iron uptake is only observed in BFR with the non-haem-iron fully depleted. Once the protein has undergone one round of the oxidative iron uptake cycle with 48 Fe(II) ions per BFR molecule then the fast oxidation phase (phase 2) can no longer be observed, indicating that the dinuclear centre is blocked [44]. Thus a Combined Mechanism may be operative which contains features of both the *Linked* and *Sequential Transfer* mechanisms.

Figure 13(c) illustrates one form of a *Combined Mechanism* in which Fe(II) enters the central cavity of the BFR via a route other than the ferroxidase centre, perhaps one of the inter-subunit channels of the protein coat, as seems likely for animal ferritin (see Section 3). Core formation is initiated at Fe(III) ions that had originally passed through the ferroxidase centre. Alternatively, it could be that an incoming Fe(II) joins the incumbent Fe(III) in the ferroxidase centre with electron transfer from an Fe(II) in the central cavity reducing the Fe(III) at the ferroxidase centre, thereby recreating a di-Fe(II) ferroxidase centre ready for reaction with O_2.

In all these suggestions the di-iron site plays a central role but two points need to be borne in mind. Firstly, in vivo BFR forms a largely Fe(III) core under strictly anaerobic conditions so the presence of external O_2 is not essential for natural core formation [20]. Secondly, variant forms of BFR with altered di-iron centres failed to exhibit ferroxidase activity as defined by a rapid burst of phase 2 Fe(II) oxidation, but they did produce an Fe(III) core as proved by the 340 nm absorbance (Fig. 1) [43]. Thus it is possible that the di-iron centre does not have a direct role in Fe(II) uptake of the type envisaged by Fig. 13, though there is no other experimental evidence pointing to another function.

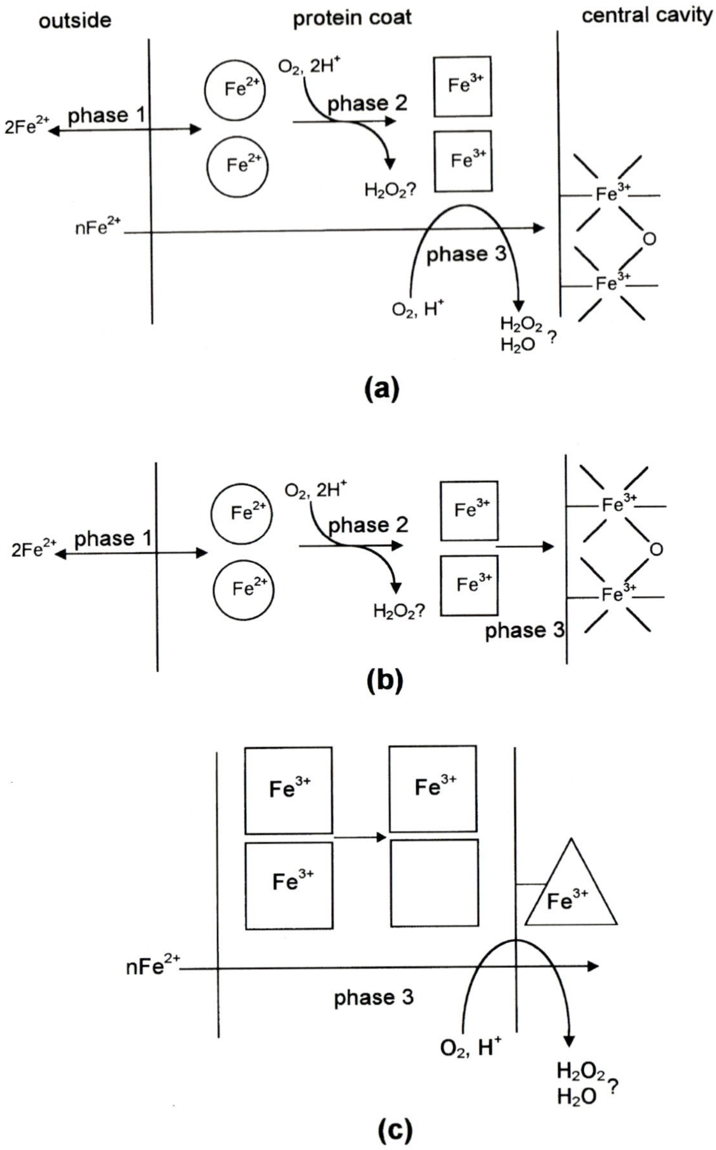

Fig. 13. Possible mechanisms of oxidative iron uptake into *E. coli* BFR involving the ferroxidase centre. **a** The *Linked Transfer* model in which phase 1 of the kinetic profile is binding of Fe(II) to the vacant ferroxidase centre, phase 2 is oxidation of the Fe(II) at the ferroxidase centre, and phase 3 is formation of an Fe(III)-core by Fe(II) entering the central cavity by a route other than through the ferroxidase centre and the ferroxidase centre acting as the Fe(II)-oxidising/O_2 reduction site. **b** The *Sequential Transfer* model in which phases 1 and 2 are the same as for the *Linked Transfer* model but phase 3 now reflects the breakdown of the di-Fe(III) ferroxidase centre and the passage of the Fe(III) ions into the central cavity. **c** One version of a *Combined* model in which phases 1 and 2 are the same as for the other two models but phase 3 is more complex (see text). (Reproduced from reference 107)

8.2 Iron Release

Iron release from BFR under physiological conditions, like that from its plant and mammalian homologues, has not been rigorously studied. In part this stems from a lack of knowledge of the physiological reaction partners of the ferritins, but it is also due to an emphasis on studies of core formation rather than iron release. One point is clear: rapid release of iron from the core requires that it be in the Fe(II) form rather than Fe(III) [6–8]. Since BFR in situ appears from Mössbauer studies to contain an oxidised core [33], this implies that relatively rapid physiological release of iron requires that it be reduced prior to release. Using the in vivo iron release assay of Fig. 2 Moore et al. [40] demonstrated that the intrinsic bis-methionine coordinated haem increased the rate of iron release compared to the equivalent haem-free protein and, moreover, that the rate of iron release from the haem-free protein depended on the reduction potential of the quinol donor, in a manner that suggested electron transfer was rate-limiting, whilst the rate of iron release from the wild-type protein was not. These observations point to a role for haem in long-range electron transfer across the protein coat, a role also suggested by Watt et al. [49], but this has not yet been confirmed by studies of haem-free BFR generated by site-directed mutagenesis. However, it is consistent with the observation of Andrews et al. [58] that the haem-free Met52His and Met52Leu variants of BFR had greater amounts of core iron than the corresponding haem-containing Met31 and Met86 variants or wild-type BFR (123–330 Fe/24mer compared to 25–42 Fe/24mer) as isolated from their over-expression system. Further studies of this phenomenon are needed, particularly as the over-expression of BFR may affect intracellular iron metabolism through a number of pathways including those associated with haem production. A role for the haem in reductive iron release is also suggested by the kinetic study of Richards et al. [108]. Using stopped-flow spectrophotometry they showed that the haem iron of *A.vinelandii* BFR was reduced by $S_2O_4^{2-}$ prior to the core Fe(III), despite the difference in their reduction potentials (see Sections 4.1 and 7.7) which would lead to the core Fe(III) be preferentially reduced if kinetic factors were not important.

It is possible that iron located at the ferroxidase site would be reduced by the addition of electrons to fully oxidised holo-BFR, as this iron appears to have a higher redox potential than the haem group (see Sections 4.1 and 5.5). Therefore, given the exchange rates reported for the binding of Fe(II) at the ferroxidase site, which indicate it is in rapid equilibrium with the bulk phase outside BFR [43, 44], some of the iron bound by BFR could be lost by this route. If the ferroxidase site were to be refilled by Fe(II) moving from the core then it might serve as a conduit for the unloading of iron, though channels through the coat (see Section 3) are generally assumed to play this role.

Acknowledgements. This work was supported by grants from the BBSRC and EPSRC, which support the UEA Centre for Metalloprotein Spectroscopy and Biology via their Biomolecular Sciences Panel, and the Wellcome Trust, whom NLB and GRM thank for a Prize Fellowship and Research Leave Fellowship, respectively. We also thank Drs. Marcia Mauk (UBC Vancouver), Simon Andrews and Amyra Treffry (Sheffield) and Profs. Mike Wilson (Essex),

Grant Mauk (UBC Vancouver), John Guest and Pauline Harrison (Sheffield) for fruitful collaborations and helpful discussions.

9 References

1. Stiefel EI, Watt GD (1979) Nature 279:81–83
2. Yariv J, Kalb AJ, Sperling R, Bauminger ER, Cohen SG, Ofer S (1981) Biochem. J. 197: 171–175
3. Moore GR, Mann S, Bannister JV (1986) J Inorg Biochem 28:329–336
4. Grossman MJ, Hinton SM, Minak-Bernero V, Slaughter C, Stiefel EI (1992) Proc. Natl. Acad. Sci. USA 89:2419–2423
5. Ford GC, Harrison PM, Rice DW, Smith JMA, Treffry A, White JL, Yariv J (1984) Phil. Trans. R. Soc. Lond. B304:551–565
6. Theil EC (1983) in "Advances in Inorganic Biochemistry" (Theil EC, Eichorn GL, Marzilli LG, eds.) Elsevier, New York, 423–431
7. Theil EC, Hase T (1993) in "Iron Chelation in Plants and Soil Microorganisms", Academic Press, New York 133–156
8. Harrison PM, Arosio P (1996) Biochim. Biophys Acta 1275:161–203
9. Arosio P, Adelman TG, Drysdale JW (1978) J Biol Chem 253:4451–4458
10. Luzzago A, Arosio P, Iacobello C, Ruggeria G, Capucci L, Brocchi E, De Simone F, Gamba D, Gabri E, Levi S, Albertini A (1986) Biochim Biophys Acta 872:61–71
11. St Pierre TG, Mann S (1989) in "Biomineralization: Chemical, Biochemical Perspectives" (Mann S, Webb J, Williams RJP eds.) VCH, Weinheim, 295–344
12. Ragland M, Briat J-F, Gagnon J, Laulhere J-P. Massenet O, Theil EC (1990) J. Biol. Chem. 265:18339–18344
13. Lobreaux S, Yewdall SJ, Briat JF, Harrison PM (1992) Biochem. J. 288:931–939
14. Wade VJ, Treffry A, Laulhere J-P, Bauminger ER, Cleton MI, Mann S, Briat J-F, Harrison PM (1993) Biochim. Biophys Acta 1161:91–96
15. Barcelo F, Areán CO, Moore GR (1995) BioMetals 8:47–52
16. Frazier BA, Pfeifer JD, Russell DG, Falk P, Olsén AN, Hammar M, Westblom TU, Normark SJ (1993) J Bacteriol 175:966–972
17. Doig P, Austin JW, Trust TJ (1993) J Bacteriol 175:557–560
18. Hudson AJ, Andrews SC, Hawkins C, Williams JM, Izuhara M, Meldrum FC, Mann S, Harrison PM, Guest JR (1993) Eur J Biochem 218:985–995
19. Hempstead PD, Hudson AJ, Artymiuk PJ, Andrews SC, Banfield MJ, Guest JR, Harrison PM (1994) FEBS Lett 350:258–262
20. Ringeling PL, Davy SL, Monkara FA, Hunt C, Dickson DPE, McEwan AG, Moore GR (1994) Eur J Biochem 223:847–855
21. Frolow F, Kalb (Gilboa) AJ, Yariv J (1993) Acta Cryst D49:597–600
22. Frolow F, Kalb AJ, Yariv J (1994) Structural Biol 1:453–460
23. Cheesman MR, Thomson AJ, Greenwood C, Moore GR, Kadir F (1990) Nature 346:771–773
24. Laulhère J-P, Labouré A-M, Wuytswinkel OV, Gagnon J, Briat J-F (1992) Biochem J 281: 785–793
25. Watt GD, Frankel RB, Papaefthymiou GC, Spartalian K, Stiefel EI (1986) Biochemistry 25:4330–4336
26. Keech AM, Le Brun NE, Mauk MR, Mauk AG, Andrews SC, Guest JR, Harrison PM, Thomson AJ, Moore GR (1997) J Chem Soc Dalton submitted
27. St Pierre TG, Bell SH, Dickson DPE, Mann S, Webb J, Moore GR, Williams RJP (1986) Biochim Biophys Acta 870:127–134
28. Mann S, Bannister JV, Williams RJP (1986) J Mol Biol 188:225–232
29. Smith JMA, Quirk A, Plank RWH, Diffin FM, Ford GC, Harrison PM (1988) Biochem J 255:737–740
30. Deeb SS, Hager LP (1964) J Biol Chem 239:1024–1031

31. Bartsch RA, Kakuno T, Horio T, Kamen MD (1971) J Biol Chem 246:4489–4496
32. Andrews SC, Harrison PM, Guest JR (1989) J Bacteriol 171:3940–3947
33. Penfold CN, Ringeling PL, Davy SL, Moore GR, McEwan AG, Spiro S (1996) FEMS Microbiol Lett 139:143–148
34. Kadir FHA, Read NMK, Dickson DPE, Greenwood C, Thompson A, Moore GR (1991) J Inorg Biochem 43:753–758
35. Almirón M, Link AJ, Furlong D, Kolter R (1992) Genes Dev 6:2646–2654
36. Peña MMO, Bullerjahn GS (1995) J Biol Chem 270:22478–22482
37. Evans Jr DJ, Evans DG, Lampert HC, Nakano H (1995) Gene 153:123–127
38. Moore GR, Pettigrew GW (1990) "Cytochromes c: Evolutionary, Structural and Physicochemical Aspects." Springer-Verlag, Berlin, 478 pp
39. Cheesman MR, Greenwood C, Thomson AJ (1991) Adv Inorg Chem 36:201–255
40. Moore GR, Kadir FHA, Al-Massad FK (1992) J Inorg Biochem 47:175–181
41. Garg RP, Vargo CJ, Cui X, Kurtz Jr DM (1996) Biochemistry 35:6297–6301
42. Levi S, Luzzago A, Cesareni G, Cozzi A, Franceschinelli F, Albertini A, Arosio P (1988) J Biol Chem 263:18086–18092
43. Le Brun NE, Andrews SC, Guest JR, Harrison PM, Moore GR, Thomson AJ (1995) Biochem J 312:385–392
44. Le Brun NE, Wilson MT, Andrews SC, Harrison PM, Guest JR, Thomson AJ, Moore GR (1993) FEBS Lett 333:197–202
45. Sun S, Chasteen ND (1992) J Biol Chem 267:25160–25166
46. Sun S, Arosio P, Levi S, Chasteen ND (1993) Biochemistry 32:9362–9369
47. Boyer RF, Clark HM, LaRoche AP (1988) J Inorg Biochem 32:171–181
48. Jones T, Spencer R, Walsh, C (1978) Biochemistry 17:4011–4017
49. Watt GD, Jacobs D, Frankel RB (1988) Proc Natl Acad Sci USA 85:7457–7461
50. Watt GD, Frankel RB, Papaefthymiou GC (1985) Proc Natl Acad Sci USA 82:3640–3643
51. Lawson DM, Artymiuk PJ, Yewdall SJ, Smith JMA, Livingston JC, Treffry A, Luzzago A, Levi S, Arosio P, Cesareni G, Thomas CD, Shaw WV, Harrison PM (1991) Nature 349: 541–544
52. Trikha J, Theil EC, Allewell NM (1995) J Mol Biol 248:949–967
53. Cheesman MR, Le Brun NE, Kadir FHA, Thomson AJ, Moore GR, Andrews SC, Guest JR, Harrison PM, Smith JMA, Yewdall SJ (1993) Biochem J 292:47–56
54. Yablonski MJ, Theil EC (1992) Biochemistry 31:9680–9684
55. Treffry A, Bauminger ER, Hechel D, Hodson NW, Nowik I, Yewdall SJ, Harrison PM (1993) Biochem J 296:721–728
56. Levi S, Santambrogio P, Corsi B, Cozzi A, Arosio P (1996) Biochem J 317:467–473
57. Al-Massad FK, Kadir FHA, Moore GR (1992) Biochem J 283:177–180
58. Andrews SC, Le Brun NE, Barynin V, Thomson AJ, Moore GR, Guest JR, Harrison PM (1995) J Biol Chem 270:23266–23274
59. George GN, Richards T, Bare RE, Gea Y, Prince RC, Steifel ED, Watt GD (1993) J Am Chem Soc 115:7716–7718
60. Moore GR: Kadir FHA, Al-Massad FK, Le Brun NE, Thomson AJ, Greenwood C, Keen JN, Findlay JBC (1994) Biochem J 304:493–497
61. Cheesman MR: Kadir FHA, Al-Basseet J, Al-Massad F, Farrar J, Greenwood C, Thomson AJ, Moore GR (1992) Biochem J 286:361–367
62. McKnight J, Cheesman MR, Reed CA, Orosz RD, Thomson AJ (1991) J Chem Soc Dalton Trans 1887–1894
63. Barker PD, Nerou EP, Cheesman MR, Thomson AJ, de Oliveira P, Hill HAO (1996) Biochemistry 35:13618–13626
64. Harbury HA, Loach PA (1960) J Biol Chem 253:3640–3645
65. Mauk AG, Moore GR (1997) JBIC 2:119–125
66. Warshel A, Papazyan A, Muegge I (1997) JBIC 2:143–152
67. Kadir FHA, Moore GR (1990) FEBS Lett 271:141–143
68. Kadir FHA, AL-Massad FK, Moore GR (1992) Biochem. J. 282:867–870
69. Moore GR, Cheesman MR, Kadir FHA, Thomson AJ, Yewdell SJ, Harrison PM (1992) Biochem J 287:457–460

70. Précigoux G, Yariv J, Gallois B, Dautant A, Courseille C, D'Estaintot BL (1994) Acta Cryst D50:739–743
71. Michaux M-A, Dautant A, Gallois B, Granier T, D'Estaintot BL, Précigoux G (1996) PROTEINS: Structure, Function, Genetics 24:314–321
72. Treffry A, Hirzmann J, Yewdall SJ, Harrison PM (1992) FEBS Lett 302:108–112
73. Bauminger ER, Harrison PM, Hechel D, Nowik I, Treffry A (1991) Biochim Biophys Acta 1118:48–58
74. Bauminger ER, Harrison PM, Hechel D, Hodson NW, Nowik I, Treffry A, Yewdall SJ (1993) Biochem J 296:709–719
75. Keech AM, Le Brun NE, Wilson MT, Andrews SC, Moore GR, Thomson AJ (1997) J Biol Chem 272:422–429
76. Inglis NF, Stevenson K, Hosie AHF, Sharp JM (1994) Gene 150:205–206
77. Nordlund P, Eklund H (1995) Curr Opin Struct Biol 5:758–766
78. Logan DT, Su X-D, Åberg A, Regnström Hajdu J, Eklund H, Nordlund P (1996) Structure 4:1053–1064
79. Rosenzweig AC, Frederick CA, Lippard SJ, Nordlund P (1993) Nature 366:537–543
80. Rosenzweig AC, Nordlund P, Takahara PM, Frederick CA, Lippard SJ (1995) Chemistry, Biology 2:409–418
81. Fox BG, Shanklin J, Ai J, Loehr TM, Sanders-Loehr J (1994) Biochemistry 33: 12776–12786
82. Feig AL, Lippard SJ (1994) Chem Rev 94:759–805
83. Atta M, Nordlund P, Âberg A, Eklund H, Fontecave M (1992) J Biol Chem 267: 20682–20688
84. Le Brun NE, Keech AM, Mauk MR, Mauk AG, Andrews SC, Thomson AJ, Moore GR (1996) FEBS Lett 397:159–163
85. da Silva JJRF, Williams RJP (1991) The Biological Chemistry of the Elements. Clarendon Press, Oxford, pp 561
86. Zhang J-H, Kurtz Jr DM (1992) Proc Natl Acad Sci USA 89:7065–7069
87. Mann S, Williams JM, Treffry A, Harrison PM (1987) J Mol Biol 198:405–416
88. Towe KM, Bradley WF (1967) J Colloid Interface Sci 24:384–392
89. Towe KM (1981) J Biol Chem 256:9377–9378
90. Powell AK (1997) Structure and Bonding 88:1–38
91. Eggleton RA, Fitzpatrick RW (1990) Clays Clay Miner 38:335
92. Rohrer JS, Islam QT, Watt GD, Sayers DE, Theil EC (1990) Biochemistry 29:259–264
93. Mackle P, Garner CD, Ward RJ, Peters TJ (1991) Biochim Biophys Acta 1115:145–150
94. Friedman JR, Sarachik MP, Tejada J, Ziolo R (1996) Phys Rev Lett 76:3830
95. Hernandez JM, (1996) Europhys Lett 35:301
96. Thomas L, Lionti F, Ballou R, Gatteschi D, Sessoli R, Barbara, B (1996) Nature 383:145-147
97. Mørup S, Dumesic JA, Topsøe H (1980) in "Applications of Mössbauer Spectroscopy" (Cohen RL ed.) Academic Press, New York, 2, 1–53
98. Bell SH, Weir MP, Dickson DPE, Gibson JF, Sharp GA, Peters TJ (1984) Biochim Biohys Acta 787:227–236
99. Le Brun NE, Moore GR, Thomson AJ (1995) Mol Physics 85:1061–1068
100. Granick S (1945) Chem Review 38:379 –403
101. Bauminger ER, Cohen SG, Dickson DPE, Levy A, Ofer S, Yariv J (1980) Biochim Biophys Acta 623:237–242
102. Matzanke BF, Müller GI, Bill E, Trautwein AX (1989) Eur J Biochem 183:371–379
103. Le Brun NE (1993) PhD Thesis, UEA, Norwich, UK
104. Weir MP, Peters TJ, Gibson JF (1985) Biochim Biophys Acta 828:298–305
105. Deighton N, Abu-Raqabah A, Rowlan IJ, Symons MCR, Peters TJ, Ward RJ (1991) J Chem Soc Faraday Trans 87:3193–3197
106. Webb J, Gray HB (1974) Biochim Biophys Acta 352:224–229
107. Thomson AJ, Le Brun NE, Keech AM, Andrews SC, Moore GR (1996) Biochem Soc Trans 25:96–101
108. Richards TD, Pitts KR, Watt GD (1996) J Inorg Biochem 61:1–13

Ribonucleotide Reductases – A Group of Enzymes with Different Metallosites and a Similar Reaction Mechanism

B.-M. Sjöberg

Department of Molecular Biology, Stockholm University, S-10691 Stockholm, Sweden.
E-mail: bitte@molbio.su.se

Ribonucleotide reductases catalyze an essential reaction in all living cells – the production of deoxyribonucleotide precursors for DNA synthesis. An intricate allosteric regulation enables one and the same enzymic component to accept both purine and pyrimidine ribonucleotide substrates and to furnish growing cells with balanced deoxyribonucleotide pools. The reaction, which is based on radical chemistry, is common to all ribonucleotide reductases, whereas the mode of furnishing the radical differs and is currently a basis for distinguishing at least three different classes of ribonucleotide reductases. The radical generating components, although of seemingly different evolutionary origin, all include transition metal binding sites. Studies on the ribonucleotide reductase family highlight exciting mechanistic and evolutionary implications of far-reaching significance.

Keywords: ribonucleotide reductase, diiron-oxo, iron-sulfur, tyrosyl radical, glycyl radical, S-adenosylmethionine

Abbreviations

ENDOR electron nuclear double resonance
EPR electron paramagnetic resonance
NMR nuclear magnetic resonance
Pfl pyruvate formate lyase
adoMet *S*-adenosylmethionine

1 Introduction

Ribonucleotide reductase denotes a family of enzymes having diverse composition and cofactor requirements, yet with a common reaction mechanism. The reaction catalyzed is the reduction of ribonucleotides to deoxyribonucleotides, essential to all living cells. By providing a balanced supply of building blocks for DNA synthesis the enzyme has a pivotal role in evolution of life on earth as well as present-day control of cell proliferation. Some severe illnesses can be related to malfunction of ribonucleotide reductase mediated via unbalanced deoxynucleotide pools [1, 2], and mutations in the enzyme proper lead to drastic disturbances in deoxynucleotide pools in ex vivo studies [3]. In addition, a variety of antiproliferative drugs directed against prokaryotic, mammalian, and viral ribonucleotide reductases have been developed during the last decade [4].

The enzymatic pathway catalyzed by ribonucleotide reductase was first discovered and characterized in the enteric bacterium *Escherichia coli* in the 1950s by Reichard and coworkers (for historical reviews see [5, 6]). This was the starting point for a successful new research field that has flourished for more than three decades and constantly offers new challenges to evolutionary, biomedical and metabolic research areas as well as horizons opened onto the fields of radical chemistry, electron transfer mechanism and metal-mediated catalysis.

The original characterization of the prototypic *E. coli* ribonucleotide reductase (called class I) was paralleled by the isolation and characterization of another prototypic enzyme (called class II) from *Lactobacillus leichmannii* [7]. Several other prokaryotes were found to have class II enzymes [8], whereas mammalian enzymes and most ribonucleotide reductases isolated from eukaryotic microorganisms belonged to class I. A useful classification method was the hydroxyurea sensitivity of class I enzymes, and the vitamin B_{12} requirement of the class II enzymes. As will be shown below, a hydroxyurea sensitive tyrosyl radical is at the heart of class I enzymes, and an adenosylcobalamin cofactor at the center of class II enzymes. In the 1980s Follmann and coworkers reported isolation and characterization of a manganese-dependent ribonucleotide reductase from *Corynebacterium* (previously *Brevibacterium*) *ammoniagenes* [9, 10], and it was realized that perhaps there were other classes than the two originally discovered. It is no exaggeration to say that the ribonucleotide reductase field has exploded with new discoveries, elegantly exemplified by the isolation and characterization of yet another class of the enzyme isolated from anaerobically grown *E. coli* (below called class III) [11, 12]. Last year's reports of complete genome sequences for an archaebacterial species [13] and cloning of ribonucleotide reductase sequences from two other archaebacteria [14, 15] are beyond doubt milestones in the ribonucleotide reductase tale as the latter two show sequence similarities to all hitherto well-characterized classes (I–III).

Have the last few years' discoveries of so many new variants of the enzyme ribonucleotide reductase made the field diverge? On the one hand yes, because the repertoire of Nature's different solutions to the production of the deoxyribonucleotides needed for DNA synthesis has expanded. In another sense no, because the addition of many new ribonucleotide reductase sequences hints at a common evolutionary origin and highlights more clearly the minimum requirements for a functional ribonucleotide reductase.

Based on more than two decades of biochemical results and the known three dimensional structure of the *E. coli* class I enzyme components, the following common features can be found: a) a quite rigidly built portion of the active site that binds the ribose moiety of the ribonucleotide to allow its reduction by redox active cysteines, b) a structurally more flexible part of the active site that accommodates the purine or pyrimidine ring of the substrate, and that responds to the allosteric regulation mediated by binding of nucleoside triphosphates at specific effector binding sites, and c) the most intriguing aspect of all – the initiation of catalysis by radical chemistry. For most ribonucleotide reductases, the radical chemistry is enabled by storage of a stable protein-derived free radical at some distance from the active site, which conceivably interacts with a metallosite, e.g. a diferric iron cluster in class I enzymes [16–18], an iron sulfur cluster in class III enzymes [19] or a manganese site in a class I variant [10, 20]. In class II enzymes the radical chemistry is initiated by splitting of a B_{12}-cofactor [7, 21]. This review will first deal with descriptions of the common features of the reaction mechanism of ribonucleotide reductases (Section 2), followed by the prototypic characteristics of the different classes (Section 3) and then will describe the different radical generating metal sites (Section 4).

2
Reaction Mechanism

The reaction catalyzed by ribonucleotide reductase is the reduction of the ribose moiety of a nucleoside di- or triphosphate to the corresponding 2′-deoxyribonucleotide (Fig. 1). I will describe the steps of the reaction mechanism as deduced for the class I reductase from *E. coli,* as this mechanism relates to the richest biochemical and genetic data, as well as structural knowledge. It is be-

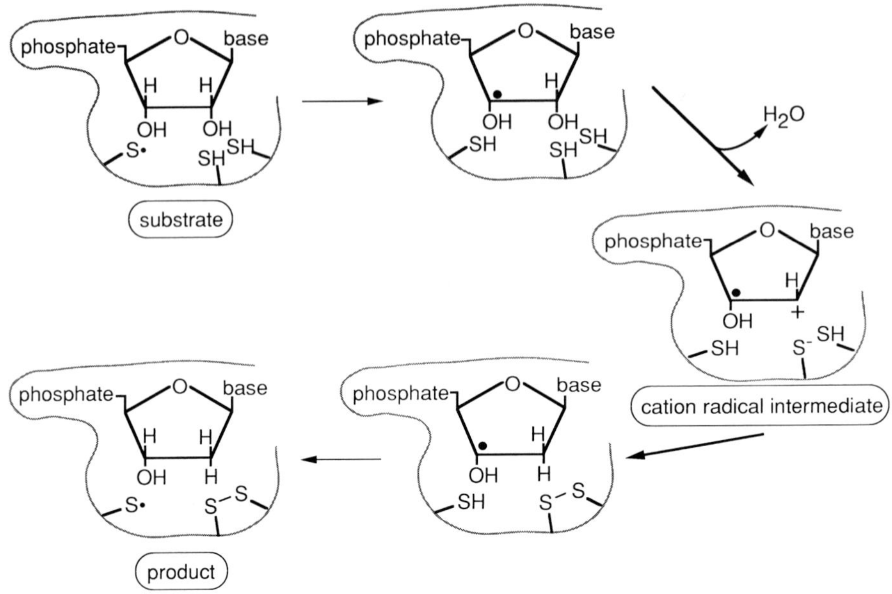

Fig. 1. Proposed general reaction mechanism

lieved that the crucial steps are common to all ribonucleotide reductases; important differences will be specified below (Chapter 3).

In stark contrast to the overall two-electron reduction, the initial step of the reaction is a one-electron oxidation of the substrate by a hitherto unobserved, transiently formed cysteinyl radical (presumably at Cys-439 in *E. coli* numbering [22, 23]), which abstracts the 3′-H atom [24]. It is believed that this oxidation facilitates leaving of the 2′-hydroxyl group, most likely protonated by one of the nearby redox active cysteines (Cys-225 in *E. coli* numbering), to generate a substrate radical cation. The radical cation is then reduced by the redox active cysteine pair (Cys-225 and Cys-462 in *E. coli* numbering), and the reaction cycle is completed by return of the abstracted 3′-H atom and release of the deoxyribonucleotide product [25].

No direct observation of a substrate radical has yet been reported, but several indirect pieces of evidence co-operate to support the suggested reaction mechanism. a) 3′-H atom abstraction is not rate limiting, but Stubbe and coworkers showed in the 1980s that after reaction 3′-tritium was retained to a greater extent in substrate than in product, implying an isotope effect in breaking of the C-H bond [24, 26]. b) Use of substrate analogues with either N_3- or SH- substitutions in the 2′-position are suicidal k_{cat} inhibitors that, after decomposition of the substrate analogue, produce transient radicals located on N or S adducts at the nearby Cys-225 (in *E. coli* numbering) [27–30]. c) A genetically engineered reductase, Cys-225 to Ser, also works as a suicidal k_{cat} analo-

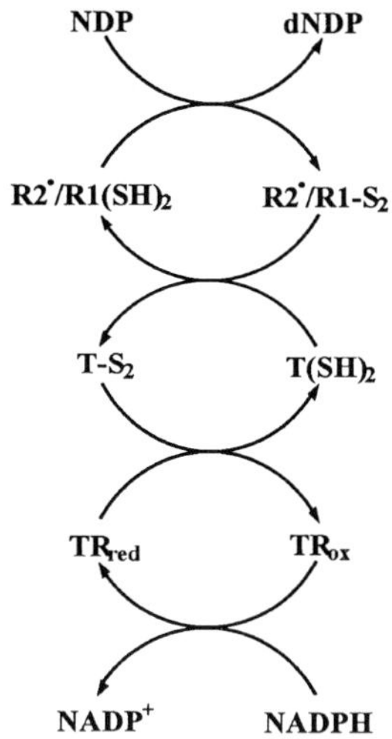

Fig. 2. Components of the ribonucleotide reductase system

gue and autotruncates at Ser-225 during catalysis, a reaction presumably preceded by formation of a transient radical at that residue [31].

One turnover cycle leaves behind in the active site an oxidized cysteine pair that needs to be reduced prior to the next turnover cycle. This reactivation of the enzyme can be performed by one of several enzyme systems: the thioredoxin system, consisting of thioredoxin, thioredoxin reductase and NADPH (Fig. 2), or the glutaredoxin system, consisting of glutaredoxin, glutathione, glutathione reductase and NADPH [32, 33]. Thioredoxin or glutaredoxin, which are the immediate reductants, interact with the conceivably flexible C-terminus of the enzyme which contains a redox active cysteine pair, usually at positions –5 and –2 from the end (Cys-754 and Cys-759 in *E. coli* numbering). It is believed that the C-terminus of the reductase can oscillate between an accessible location where it is reduced by thioredoxin or glutaredoxin, and a more crowded location, where it interacts with and reduces the disulfide formed at the active site [34]. The overall turnover rate of approximately 10 s^{-1} for the ribonucleotide reductase system is surprisingly fast for a mechanism requiring so many conformational transitions.

3
Ribonucleotide Reductase Classes

Three well-characterized classes of ribonucleotide reductase have been described so far (Table 1) [35, 36]. Classes I and II have obvious regions of amino acid sequence similarities (see below) whereas the connection between class III and the other two is less obvious. Last year's isolation and partial enzymatic characterization of ribonucleotide reductases from Archae species [13–15] have however changed the picture drastically as two examples of extensive patch-

Table 1. Characteristics of different classes of ribonucleotide reductases

Type of class	Polypeptide composition	Gene nomenclatur	Stable radical/ cofactor	Metal site	Substrate	Reductant	K_{cat} (s^{-1})
Class I	$\alpha_2\beta_2$	nrdA/nrdE nrdB/nrdF	Tyr	Diiron-oxo	NDP	Thioredoxin Glutaredoxin X-redoxin	4–11
Class II	α or α_2	–	AdoCbl	Cobalamin	NDP/NTP	Thioredoxin	1.6
Class III	$\alpha_2\beta_2$	nrdD nrdG	Gly/AdoMet	Iron-sulfur	NTP	Formate	3

work from all three classes have been described (see below). This definitely speaks in favor of a common evolutionary origin for all ribonucleotide reductases as regards the substrate binding component (α polypeptide) including allosteric regulatory regions. In contrast, the cofactor or component providing the oxidized radical needed for initiation of catalysis differs between different classes (Table 1), and has distinctly different evolutionary relationships.

3.1 Class I

Class I currently constitutes the most extensive class, with more than 25 completely sequenced genes and at least 2 subclasses, Ia and Ib. It comprises ribonucleotide reductases from all types of eukaryotes, several eukaryotic viruses, some prokaryotes and some bacteriophages. The description of the class I ribonucleoside diphosphate reductase will concentrate on the *E. coli* Ia subclass (encoded by the *nrdAB* genes), which has been most well-characterized.

Two α-polypeptides (encoded by the *nrdA* gene) constitute the homodimeric R1 protein of 761 residues. Its high resolution three-dimensional structure (Fig. 3a) was solved in 1994 by Uhlin and Eklund [34, 37]. The active site occupies one side of a 10-stranded β/α-barrel, comprised of two antiparallel half-barrels, which is pierced by a long loop (Fig. 3a). The 3 invariant redox active cysteine residues of the active site are on two of the barrel β-strands (Cys-225, Cys-462) and at the tip of the piercing loop (Cys-439). A combination of structural and biochemical data has enabled identification of one of the two allosteric nucleotide binding sites (the so called specificity site) [38a] close to the polypeptide interaction area in R1, and also close to the holoenzyme interaction area. Biochemical data from several mammalian representatives of class I [3; M. Meuth, personal communication; L. Thelander, personal communication] also point to a location for the other allosteric site (the so called activity site) in the *N*-terminal domain of protein R1. Recent structural studies confirm this interpretation [38b].

Two β-polypeptides (encoded by the *nrdB* gene) constitute the homodimeric R2 protein of 375 residues. Its high resolution structure (Fig. 3a) was solved in 1990 by Nordlund and Eklund [17, 18]. Recently the structure of mouse R2 was

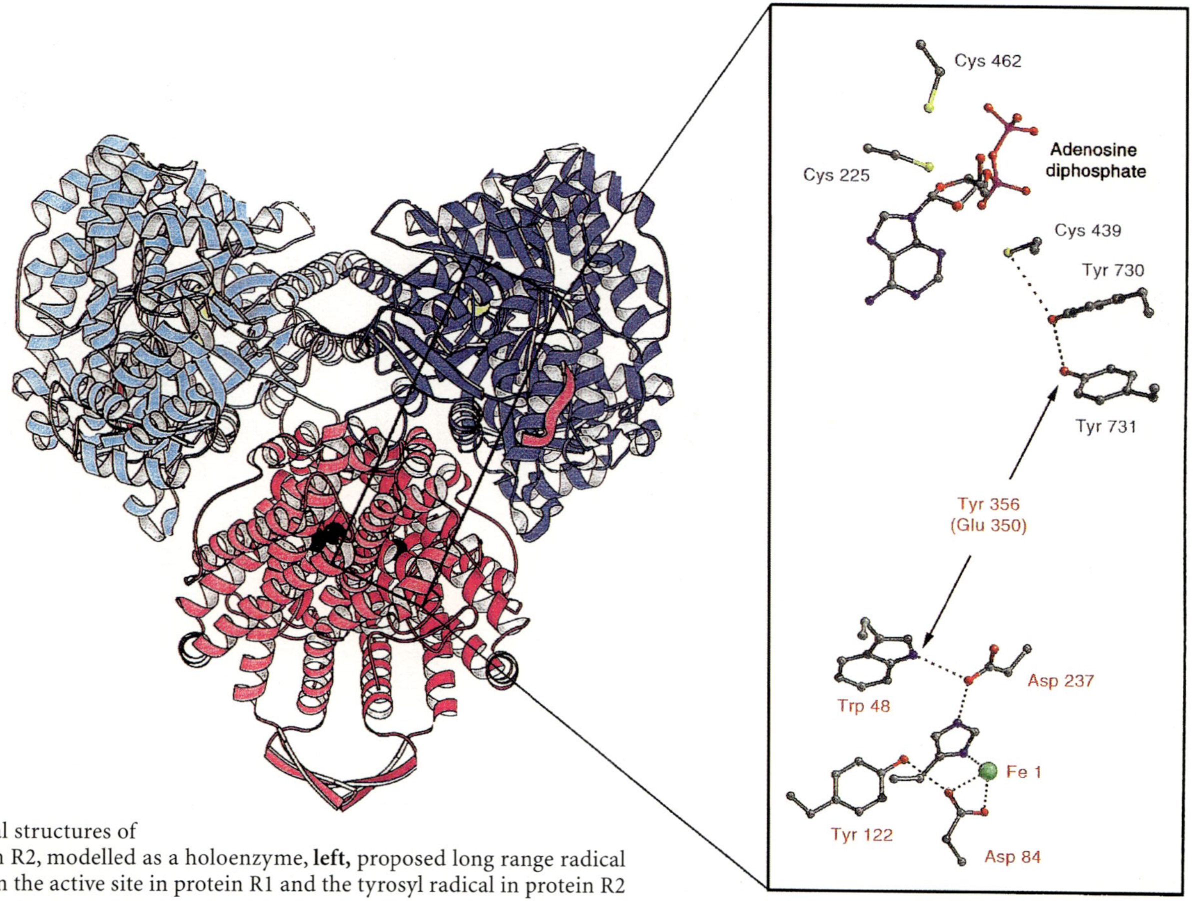

Fig. 3. Three-dimensional structures of
right, protein R1, protein R2, modelled as a holoenzyme, **left,** proposed long range radical transfer pathway between the active site in protein R1 and the tyrosyl radical in protein R2

also solved [39, 40]. The R2 protein has attracted well-deserved attention as it was the first protein discovered to carry a stable tyrosyl radical [41], now known to reside at the invariant Tyr-122 [42] close to a ferric diiron-oxo site [17, 18] (Fig. 4).

The tyrosyl radical is believed to initiate catalysis in the R1:R2 (or $\alpha_2\beta_2$) holoenzyme via a long range hydrogen bonded pathway (Fig. 3b) [43] built by at least 6 invariant residues in R2 (Y122, the iron ligands D84 and H118, and D237, W48, Y356, and perhaps E350) and 3 invariant residues in R1 (Y730, Y731, and C439). Site-directed mutagenesis of all these residues in *E. coli* R1 and R2 [22, 42, 44–46] and some of these residues in mouse R2 [47; L. Thelander, personal communication] have demonstrated their indispensability during catalytic turnover.

The Ib subclass (encoded by the *nrdEF* genes) comprises the two homodimeric components $R1_E$ and $R2_F$ [48]. Class Ib has been found in several prokaryotes along with genes for class Ia [49, 50], but in the recently completed genome of *Mycoplasma genitalium* class Ib is the only ribonucleotide reductase locus found [51]. In *E. coli* the *nrdEF* locus is not translated in sufficient amount to support growth under normal laboratory conditions [48]. The similarity between class Ia and class Ib is extensive as demonstrated in Fig. 5. The eukaryotic class I enzymes can currently be assigned to either of these prokaryotic subclasses. Preliminary genetic and biochemical data indicate that the manganese-dependent ribonucleotide reductase from *Corynebacterium ammoniagenes* [10] also belongs to class Ib [52].

Class Ib enzymes usually require their own low molecular weight reduction protein for catalytic turnover. This specific "redoxin" is usually encoded by the *nrdH* gene in close proximity of the *nrdEF* genes. The NrdH product is more closely related to glutaredoxin than thioredoxin in amino acid sequence comparisons, but lacks the glutathione binding motif [50].

3.2
Class II

Class II contains only one representative that has been both extensively biochemically characterized and completely sequenced: the monomeric α-polypeptide equivalent of *L. leichmannii* [53]. However an open reading frame from the mycobacteriophage L5 displays 36% identity (56% similarity) to the *L. leichmannii* α-polypeptide (Fig. 5) [54]. A requirement for adenosylcobalamin as cofactor has hitherto been used as a common character of class II, and B_{12}-dependent ribonucleotide reduction has been demonstrated for a variety of prokaryotes and some eukaryotic microorganisms [8, 55]. The near future will beyond doubt clarify the phylogenetic relationship between different B_{12}-dependent ribonucleotide reductases, as sequencing projects are ongoing for both cyanobacterial species [56; F. Gleason, personal communication] and *Corynebacterium nephridii* [M. Karlsson and H.P.C. Hogenkamp, personal communication]. Two recently purified archaebacterial ribonucleotide reductases are also B_{12}-dependent. As they have in fact more sequence similarities to representatives of class I than class II (Fig. 5) they will be dealt with separately (see below).

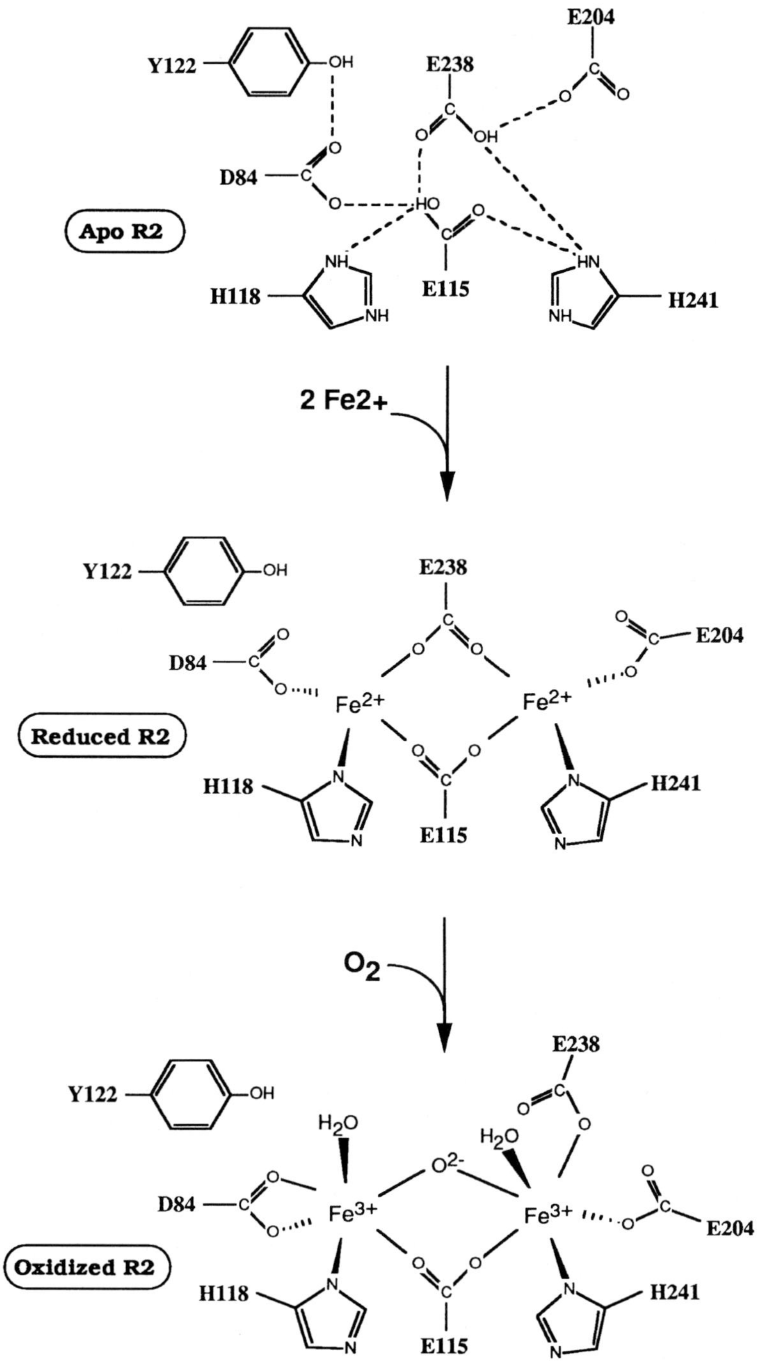

Fig. 4. Structural rearrangements occurring during reconstitution of the diferric-oxo site in protein R2

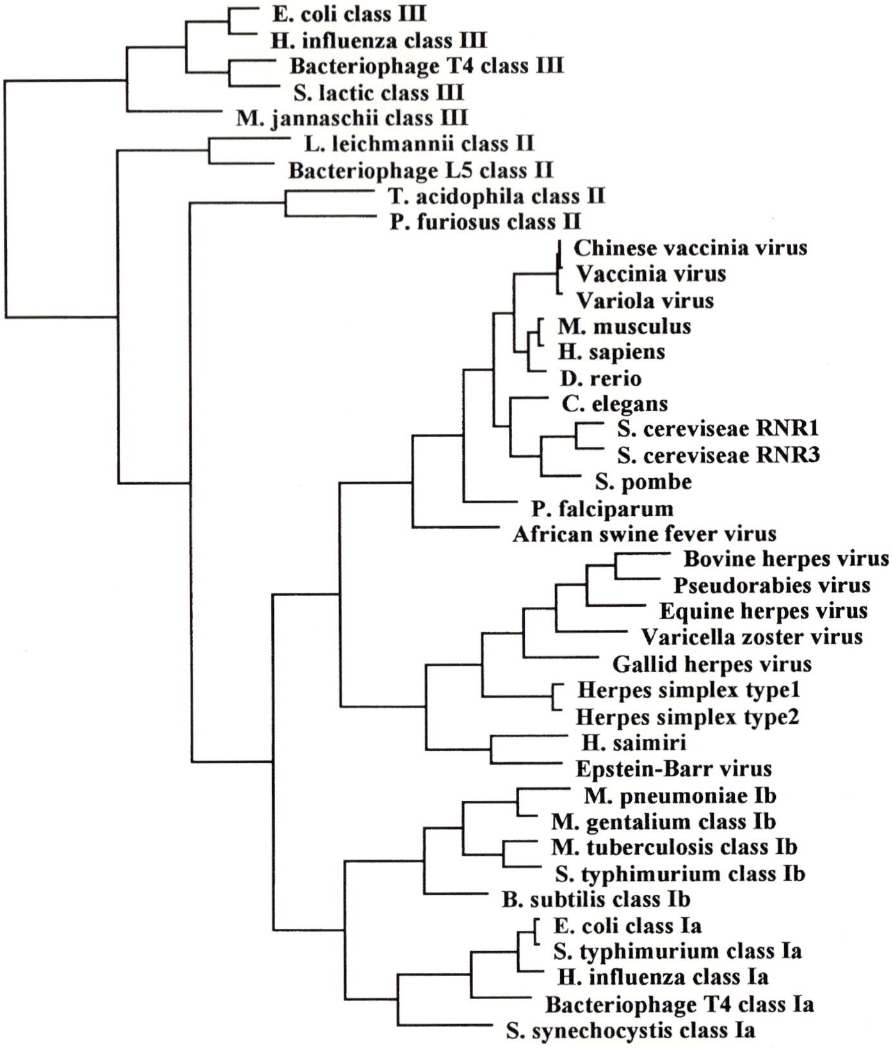

Fig. 5. Phylogenetic comparison of deduced amino acid sequences for ribonucleotide reductase α-polypeptides of class I, II, III sequences. Sequences were aligned with Pileup of the GCG program package, readjusted manually to align the equivalent of *E. coli* R1 Cys-225 of all sequences, except *L. leichmannii*, bacteriophage L5, the equivalent of *E. coli* R1 Cys-439 of all sequences, the equivalent of *E. coli* R1 Cys-462 of all sequences, except class III sequences, the equivalent of *E. coli* R1 Cys-754, Cys-759 of all sequences except *T. acidophila*. The PAUP program was used to compute a heuristic phylogenetic tree based on the aligned sequences, with *E. coli* Pfl as an outgroup

The B_{12}-dependent ribonucleotide reductase from *L. leichmannii* is a monomeric protein of 738 residues [53], whereas several other purified B_{12}-dependent ribonucleotide reductases have proved to be homodimers [55]. Three cysteine residues in *L. leichmannii* ribonucleoside triphosphate reductase (Cys-119, Cys-408, and Cys-419, corresponding mechanistically to C225, C439 and C462 in *E. coli* class Ia) are essential for catalysis [57], and believed to build an active site with structural and functional similarities to the R1 active site. The C-terminal part of the protein contains C731/C736 and a motif with extensive similarities to that of the *E. coli* C-terminal cysteines (C754 and C759) known to interact with thioredoxin and glutaredoxin. As the *L. leichmannii* enzyme depends on thioredoxin for catalytic turnover, it is conceivable that a similar mechanism to that proposed above for reactivation of protein R1 of class I also applies to the class II enzyme.

3.3
Class III

Class III includes completed sequences from 5 prokaryotic species (Fig. 5). This is an oxygen-sensitive enzyme found in some facultative anaerobes. The description of class III enzyme will be based on biochemical and genetic data from *E. coli* ribonucleoside triphosphate reductase [12, 58] supplemented with some data from the bacteriophage T4 enzyme [59–61].

The class III enzyme is encoded by the *nrdDG* genes, containing information for the α- and β-polypeptides, respectively. The *E. coli* α-polypeptide contains 712 residues and forms the homodimeric NrdD product. This component binds substrate nucleotides as well as allosteric effector nucleotides [62]. It also harbors a stable glycyl radical at Gly-680 in the *E. coli* enzyme [12] and at Gly-581 in the T4 enzyme [61], which is essential for enzyme activity and considered to be the entity that initiates catalysis. The 155-residue β-polypeptide forms the homodimeric NrdG component, which contains an iron-sulfur cluster [19]. It is believed that the holoenzyme is a complex of the two homodimeric NrdD:NrdG components [19, 61], and that the glycyl radical in the NrdD component is generated by homolytic cleavage of *S*-adenosylmethionine by the NrdG component in a reaction where an external electron is provided by flavodoxin, flavodoxin reductase and NADPH [63, 64].

There are seven conserved cysteine residues in the NrdD polypeptide (cf. Fig. 7): two single cysteines in the *N*-terminal half of the polypeptide chain (Cys-79 and Cys-290 in T4 NrdD), two –CXXC-motifs towards the C-terminal end (Cys- 543, Cys-546, Cys-561 and Cys-564 in T4 NrdD), and one cysteine adjacent to the glycyl radical residue (Cys-679 in *E. coli* NrdD). Mutational studies in the bacterial enzyme showed that a mutant C679S protein still retained some enzyme activity, thus identifying this residues as dispensable [12]. Preliminary studies on mutational changes in T4 NrdD have identified C79S and C290S as enzymatically inactive but capable of glycyl radical formation, and C543S, C546S, C561S and C564S as defective in glycyl radical formation [J. Andersson, unpublished]. These results, as yet preliminary, indicate that the class III enzyme may also conform to a reaction mechanism involving redox active cy-

A. Ribonucleotide reductase NrdD

```
                   248                                                                            324
Bacteriophage T4   DPNYDIKQLA ...LECASKR MYPDIISAKN NKAITGSSVP VSPMGCRSFL SVWKDSTGNE ILDGRNNLGV VTLNLPRIAL
E. coli            DPNYDIKQLA ...LECASKR MYPDILNYDQ VVKVTGSFKT ..PMGCRSFL GVWENENGEQ IHDGRNNLGV ISLNLPRIAL
H. influenza       DPNYDIKQLA ...LECASKR MYPDILNYDQ VVKVTGSFKA ..PMGCRSFL GAYE.EKGHE IHDGRNNLGV VSLNLPRIAL
S. lactis          TPNYDIKELA ...LECSTKR MYPDILSYDK IVELTGSFKA ..SMGCRSFL QGWKDANGND VTAGRNNLGV VTVNLPRIAL
M. jannaschii      ELMYKIHQLS AKFGIPYFIN MLPDWQVTNT .......... .NAMGCRTRL SGNWTGDAEI DTLRTGNMQW YSLNLPRIAY
E. coli Pfl        AAKVSIDTSS ..LQYENDDL MRPDFNNDDY .......... .AIACCVSPM IVGK...QMQ FFGARANLAK TMLYAINGGV
     Consensus     ---y-I--l- ---------- M-PD------ ----t----- --mgCr--l- ---------- ------N--- ---nlpria-
                                                                  *
```

```
                   536                                                                                        606
Bacteriophage T4   GVNMPVDKCF TCGSTHEMTP TENGFVCSIC GETD...... ........PK KMNTIRRTCG YLGNPNERGF NLGKNKEIMH RVKHQ*.... ..........
E. coli            GTNTPIDECY ECGFTGEFEC TSKGFTCPKC GNHD...... ........AS RVSVTRRVCG YLGSPDARPF NAGKQEEVKR RVKHLGNGQI G*........
H. influenza       GTNTPIDECY ECGFTGEFEC TSKGFVCPKC GNHD...... ........ST KVSVTRRVCG YLGSPDARPF NAGKQEEVKR RVKHL*.... ..........
S. lactis          GTNAPIDHCY ACGFEGDFTP TERGFKCPQC GNDD...... ........PK TCDVVKRTCG YLGNPQARPM VHGRHKEISS RVKHMNGSVG ALNDGNLIDS
M. jannaschii      TYTKNLSVCN RCGI....SM GGLRDRCINC G......... ........SE DVAKFSRITG YLQNISNW.. NRAKQKELED RKLPRI*... ..........
E. coli Pfl        GKDDEVRKTN LAGLMDGYFH HEASIEGGQH LNVNVMNREM LLDAMENPEK YPQLTIRVSG YAVRFN..SL TKEQQQDVIT RTFTQSM*.. ..........
     Consensus     --------c- -cG------- ------c--c g--------- ---------- ------R--G Yl-------- ------e--- R--------- ----------
                                                                                  ↑
```

B. Ribonucleotide reductase NrdG

```
                                    1                                               43
Bacteriophage T4   ..... .......... MNYDRIY.PC DFVNGPGCRV ......VLFV TGCLHKCEGC YNRSTWNARN
E. coli            ..... .........M VNYHQYY.PV DIVNGPGTRC ......TLFV SGCVHECPGC YNKSTWRVNS
H. influenza       ..... .......... MNYLQYY.PT DVINGEGTRC ......TLFV SGCTHACKGC YNQKSWSFSA
S. lactis          MNNPK PGEWRADELS QNYIADYKPF NFVDGEGVRC ......SLYV SGCMFHCEGC YNQATWSFRY
M. jannaschii      ..... .......... ...MKALVS. GIVDLSTIDY PKKASAVIFL YGCNMKCPYC HNLK.FMLEH
E. coli Pfl act.   ..... .......MSV IGRIHSFESC GTVDGPGIRF ......ITFF QGCLMRCLYC HNRDTWDTHG
E. coli FNR        ..... .......... .......... .......MIP EKRIIRRIQS GGCAIHCQDC SISQLCIPFT
     Consensus     ----- ---------- ---------- ---------- ---------- -GC---C--C -n--------
                                                                   *   *  *
```

Fig. 6. Schematized amino acid sequence comparison between components of class III ribonucleotide reductase, the Pfl family. Only selected regions have been highlighted. **a** The NrdD, Pfl components. An *arrow* denotes the site of the stable glycyl radical, an *asterisk* the site of a transient cysteinyl radical formed in Pfl during catalysis. This cysteine residue is also conserved in the NrdD family. **b** the NrdG, Pfl activase, FNR components. An *asterisk* denotes potential iron ligands in NdrdG, Pfl activase; the known iron ligands in FNR are C20, C23 (highlighted in the figure) plus C29, C122 [120]. The consensus line shows residues conserved in all aligned species (upper case), residues conserved only in the class III family (lower case)

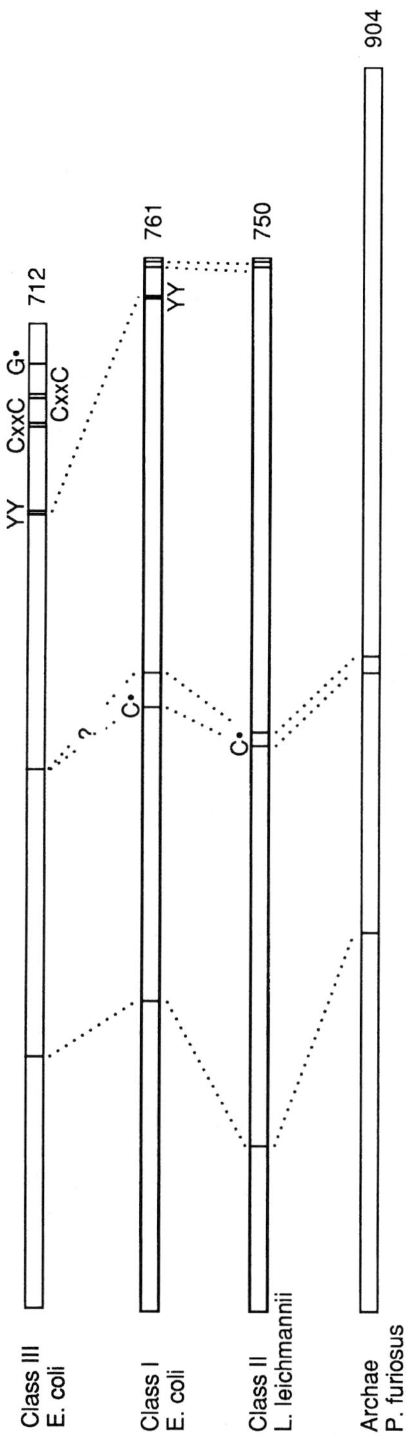

Fig. 7. Schematic representation of conserved cysteines in the different classes of ribonucleotide reductase

steines, and that the two more C-terminal -CXXC- motifs may be involved in radical generation, perhaps by interacting with the NrdG components and/or its cofactors. The reductive reactivation of the class III enzyme does not involve a thioredoxin or glutaredoxin system, but can be performed by formate, which is oxidized to carbon dioxide [65].

There are several compelling similarities between the anaerobic class III ribonucleotide reductase and the glycolytic enzyme pyruvate formate lyase (Pfl) [66–68] suggesting a common evolutionary origin: Both require homolytic cleavage of *S*-adenosylmethionine promoted by a smaller component (NrdG and the so called Pfl activase, respectively) and the flavodoxin system to generate a glycyl radical essential for activity in a larger component (NrdD and the Pfl proper, respectively). In addition, the amino acid sequence in the vicinity of the glycyl radical is conserved between NrdD and Pfl (Fig. 6), as well as the amino acid sequence in the *N*-terminal part of NrdG and Pfl activase (see below, cf. Fig. 6). Interestingly, Pfl catalysis has been shown to involve a cysteinyl radical, presumably at Cys-419 [69, 70], with some sequence similarity to the region around invariant Cys-290 in T4 NrdD (Fig. 6). Another intriguing fact is the involvement of formate in both systems; in the class III enzyme to reactivate the oxidized NrdD component during turnover, in Pfl as one of the products of the reaction.

3.4 Archaebacterial Ribonucleotide Reductases

Recently three archaebacterial ribonucleotide reductases were sequenced, from the thermophilic strict anaerobe *Methanococcus jannaschii* [13], the thermophile *Thermoplasma acidophila* [15], and the hyperthermophilic strict anaerobe *Pyrococcus furiosus* [14]. The completely sequenced genome of *M. jannaschii* reveals the presence of a 764-residue open reading frame with 27% identity (51% similarity) to the *E. coli* NrdD protein. The genes for the other two archaebacterial reductases were isolated based on partial peptide sequences obtained from homogeneous proteins purified according to their enzymatic activities. These two enzymes are both B_{12}-requiring ribonucleoside diphosphate reductases. In addition, the deduced amino acid sequences for the *M. jannaschii* and *P. furiosus* enzymes contain each two intein sequences, i.e. self-splicing peptide sequences.

A surprising finding is that the *M. jannaschii* genome appears to lack an open reading frame corresponding to the NrdG protein. However, there is an open reading frame with 20% identity (55% similarity) to Pfl activase in the absence of an open reading frame corresponding to Pfl proper. As there are defined similarities between NrdG and Pfl activase, especially in the *N*-terminal part, comprising three invariant cysteines and an invariant asparagine it is possible that the *M. jannaschii* Pfl activase homologue is a ribonucleotide reductase class III component. Interestingly, the *M. jannaschii* genome appears to code only for a class III enzyme.

The B_{12}-dependent ribonucleotide reductases from *T. acidophila* and *P. furiosus* show 40% identity (61% similarity) to each other. In addition, they have stretches of sequence similarity to both class I and class II enzymes, and to a minor extent even to the class III family. The most interesting result deduced from the sequences is that they contain equivalents of the three redox active, catalytically essential cysteines known to be located to the active site in protein R1 of class I reductase (Fig. 7). This is an important finding that for the first time identifies evolutionary links between all hitherto sequenced classes and enables calculation of a composite phylogenetic tree for the α-polypeptide of ribonucleotide reductases (Fig. 5). It reveals that the two archaebacterial B_{12}-dependent enzymes are closer to class I than is the well-characterized B_{12}-dependent enzyme from *L. leichmannii.* It also shows that this last-mentioned representative, the related open reading frame from mycobacteriophage L5 and the class III enzymes are more divergent from the other ribonucleotide reductase representatives. These important phylogenetic milestones strongly underscore that the reaction mechanism of ribonucleotide reduction has been preserved throughout evolution.

It has been speculated that archeabacteria are more related to eukaryotes than are eubacteria, and the *M. jannaschii* genome seems to corroborate this. Its genes for the transcriptional and translational machineries as well as the replicative genes are more similar to eukaryotic than eubacterial counterparts, whereas genes for its intermediary metabolism are more similar to the eubacterial counterparts. In this context, it is highly interesting that the sole *M. jannaschii* ribonucleotide reductase belongs to a class only found in eubacteria. As the ribonucleotide reductases isolated from the other two archeabacterial species are B_{12}-dependent, it seems that also these two enzymes are more similar to eubacterial than eukaryotic counterparts. It still remains an open question whether *T. acidophila* or *P. furiosus* codes for more than one ribonucleotide reductase. However, even though several prokaryotic species have been shown to code for more than one class of ribonucleotide reductase, no organism with a B_{12}-dependent enzyme has yet been shown to encode more than one ribonucleotide reductase enzyme. At any rate, the observation that the first three characterized archeabacterial ribonucleotide reductases are more similar to eubacterial than eukaryotic enzymes implies that ribonucleotide reduction is a function developed before the divergence of the three present day branches, eubacteria, archea and eukaryotes, on the tree of life.

4 Metallosites

Whereas the commonly accepted and general mechanism for ribonucleotide reduction requires the generation at the active site of a transient radical, the generation of the radical can be achieved by more varied means, like cleavage of a cofactor and/or storage of a stable protein derived radical followed by radical transfer. It is, however, a dogma that all hitherto characterized ribonucleotide reductases also contain metal ion(s) closely linked to radical generation and in only a few types of chemical environments. These structures, as well as their chemical properties and importance in the enzyme reaction, are described in detail below.

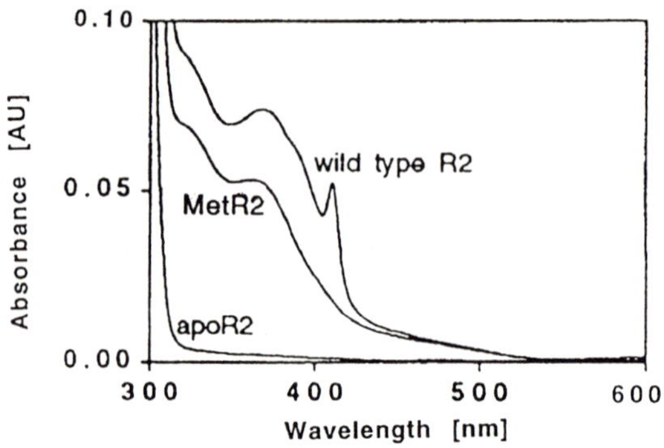

Fig. 8. Electronic absorption spectra of **a** wild type R2, **b** metR2,, **c** apoR2 of *E. coli* class Ia ribonucleotide reductase

4.1 Dinuclear Iron Sites

The diiron site of class I enzymes is by far the most well-characterized type of ribonucleotide reductase metal site. The early characterization of this site reposed to a great deal upon its compelling similarity in electronic absorption to the dinuclear iron site of hemerythrin [71], with prominent bands at 325 and 370 nm (Fig. 8) ascribed to the diferric-oxo site (see below) [16]. Later studies have shown that hemerythrin in fact constitutes its own subgroup of diiron-oxo sites with a predominance of histidine ligands, whereas protein R2 of class I ribonucleotide reductase is a typical representative of a abundant group of diiron sites with a majority of carboxylate ligands [72, 73].

The function of the diferric site in the protein R2 during catalysis, i.e.after radical generation, still remains an enigma. Nevertheless, the redox potential (see below) of the diiron site in R2 shifts to a considerably more negative value upon complex formation with protein R1 [74], and the presumed hydrogen bonded pathway suggested to operate during catalysis may involve at least one of the ferric ions (cf. Fig. 3b) [43]. In contrast, the role of the diiron site during generation of the stable tyrosyl radical is well established, and the reaction mechanism, which has been resolved in considerable detail [75, 76], includes oxygen-dependent oxidation of a diferrous site with a concomitant oxidative generation of the stable free radical at the adjacent tyrosine residue.

4.1.1 *Structure of the Diiron-oxo Site*

The three-dimensional structure of the diferric site in R2 [17, 18] consists of two irons at a distance of 3.3–3.4 Å coupled by one oxo-bridge and one carboxylate bridge (Glu-115). The dihedral angle of the Fe-O-Fe bridge is $\approx 130°$ [77]. The two irons are denoted Fe1 and Fe2 based on their vicinity to radical residue Tyr-122 (Fig. 4 and 12a). The coordination sphere of the partially distorted octahedron of Fe1 includes a bidentate carboxylate (Asp-84), an imidazole (His-118), and a water molecule, and that of octahedrally coordinated Fe2 includes two monodentate carboxylates (Glu-204, Glu-238), an imidazole (His-241) and a water molecule. All ligands of the diiron site are part of a four helix bundle, built up from two antiparallel dihelical structural motifs. The second coordination sphere includes the radical harboring side chain at Tyr-122 ca. 5.3 Å from Fe1, one carboxylate (Asp-237) and one serine (Ser-114) hydrogen bonding to His-118 and His-241, and Gln-43 hydrogen bonding to both Asp-237 and Ser-114. One prominent extended hydrogen bonded path made up of highly conserved residues (His-118, Asp-237, and Trp-48) connects Fe1 with the surface of the R2 protein that interacts with protein R1 (cf. Fig. 3). The Fe1:Y122 region is surrounded by a hydrophobic pocket consisting of conserved residues Phe-208, Phe-211, Ile-234 and non-conserved Ile-231, the first three of which have been shown to be major determinants for the stability of the tyrosyl radical [78].

Recently the 390-residue mouse R2 structure was solved to 2.3 Å resolution [40]. The subunit structure of mouse R2 is very similar to that of *E. coli* R2 and the central parts of the two structures (262 residues) can readily be superimposed with a root mean square deviation of 1.6 Å in C_α positions. A major difference is that whereas the *E. coli* R2 protein crystallized at pH 6.1 shows fully occupied diiron sites, the mouse R2 protein crystallized at pH 4.7 only shows metal related electron density in the position equivalent to Fe2. This is in accordance with the fact that the diiron site of mouse R2 is considerably less stable than that of *E. coli* R2. A comparison of the positioning of side chains in the vicinity of the iron site in mouse R2 with that of *E. coli* diferric and apoR2 strongly suggests that the coordination of the diferric sites is the same. An interesting observation is that the tyrosyl radical/diiron site in mouse R2 is accessible to solvent via a narrow hydrophobic channel, explaining its higher susceptibility to radical scavengers and metal chelators as compared to *E. coli* R2 where such a channel is lacking [79–81].

The iron site in nascent R2 is formed by binding of ferrous iron to apoR2 (Fig. 4) and the structures of the ferrous form (denoted reduced R2) and apo R2 have been solved to high resolution [82, 83]. Reduced R2 can also be generated by chemical or enzymatic reduction of ferric R2 [84–86]. The midpoint potential for the diferric/diferrous redox cycle in R2 is –115 mV, and reduction occurs as a two-electron process coupled with a protonation event [74]. As evident from figure 4, substantial conformational rearrangements are needed both upon binding of ferrous ions to apoR2 and upon oxidation of the diferrous site or reduction of the diferric site.

The structure of the empty iron site in apoR2 reveals an interesting hydrogen bonded network of the ligands-to-be by involvement of a few water molecules and protonation of some carboxylate side chains [82]. Thus the overall charge of this potentially unfavorably charged site is zero. Interestingly, this seems to be a generally preferred feature as filled iron sites in R2, e.g. the native diferric site described above and the diferrous site described below, also have zero net charges. In addition, recent structure determinations of empty iron sites in several mutant R2 proteins lacking one of the critical side chains suggest compensation for the change in overall charge by different degrees of carboxylate protonation [87].

Reduced R2 contains two diferrous ions at a distance of 3.9 Å bridged by Glu-115 and Glu-238 [83]. Each iron ion also coordinates one carboxylate and one imidazole: for Fe1 these are monodentate Asp-84 and His-118, for Fe2 monodentate Glu-204 and His-241. No water ligands are seen in the 1.7 Å electron density map.

A mixed valent form of protein R2 has not been crystallized so far. Spectroelectrochemical titration studies showed clearly that *E. coli* R2 does not stabilize a mixed valent form in solution, thereby differentiating it from most other diiron-oxo proteins studied so far [74]. However, mild chemical reduction of the mouse R2 protein produced a typical $S = {}^1/_2$ rhombic EPR signal, presumably with a hydroxo bridged mixed valent site as judged from its weak antiferromagnetic coupling [88]. In addition, mixed valent forms of *E. coli* R2 have been generated by one electron reduction of diferric R2 by diimide or hydrazine [89, 90], or low temperature ionizing radiation [91–93]. The EPR signals of the mixed-valent forms are of two types, most likely indicative of different ligand geometry at the diiron site. Low temperature reductions give rise to an axial EPR spectrum typical of an antiferromagnetically coupled $S = {}^1/_2$ mixed valent state. Annealing at increasingly higher temperatures produces initially a rhombic $S = {}^1/_2$ EPR signal at 140 K, whereas annealing at ≥160 K results in drastically changed EPR signals with *g* values of 14.8, 6.7, and 5.4, indicative of a ferromagnetically coupled $S = {}^9/_2$ mixed valent state, which disappears after annealing at 230 K, when the diferric form is regenerated. The axial signal represents in all probability a mixed valent species that still retains the diferric geometry and the oxo-bridge, whereas the rhombic signal and the $S = {}^9/_2$ states represent discrete geometrical changes and/or protonations of the ligands, including a hydroxo-bridged diiron site [94].

4.1.2
Mechanism of Tyrosyl Radical Generation

Tyrosyl radical generation occurs during dioxygen oxidation of the ferrous form of R2 according to Eq. 1:

$$\mathrm{R2(Fe^{II})_2\text{-}Y122 + O_2 + e + H^+ \rightarrow R2(Fe^{III}\text{-}O^{2-}\text{-}Fe^{III})\text{-}Y122 + H_2O} \qquad (1)$$

As Eq. 1 suggests, the μ-oxo bridge originates from dioxygen, yet it exchanges with solvent water on the minute time scale [95]. The net reduction of dioxygen to water is a four-electron process; two electrons are supplied by the metal ions, one by Tyr-122, and the fourth usually by reductants in the solvent, e.g. excess Fe^{2+}. The stable end products, Y122 and the oxo-bridged diferric site appear with rate constants of 2.5 s^{-1} at room temperature (ca. 1 s^{-1} at 4 °C) [75, 96].

The reconstitution reaction results in a deprotonated tyrosyl radical as shown by a) its high *g* value of 2.0047 [97], b) its high spin density at the phenolic oxygen and correspondingly low spin density at the adjacent C-4 (0.29 and −0.05, respectively) [98], c) its resonance Raman band at 1498 cm^{-1}, which is unperturbed by incubation in deuterated solvent [77] and d) its sharp electronic absorption band at 410 nm (Fig. 8). The distance of 5.3 Å between Tyr-122 and Fe1 excludes a direct interaction between the two, and the microwave saturation behavior of the tyrosyl radical corroborates a weak interaction with the diiron site. Several attempts at measuring the redox potential of the radical have been hampered by its inaccessibility to most reductants. The best estimate is $+1000 \pm 100$ mV [74], as expected for a tyrosyl radical and in sharp contrast to the 1100 to 1200 mV more negative potential of the adjacent diiron site.

The characteristic EPR hyperfine doublet of the stable tyrosyl radical in ribonucleotide reductase relates to the coupling between the spin density at C-1 and one of the C-β protons and is highly geometry dependent. Thus the typical EPR signals from different organisms differ in appearance due to each species' specific orientation of the aromatic ring in relation to the geometry of C-β (Fig. 9). On one extreme are the EPR signals from *E. coli,* bacteriophage T4 and mouse R2 proteins with a torsional angle for one of the β-protons of ca. 30° out-of-plane, and on the other extreme the EPR signal from *S. typhimurium* protein $R2_F$ with a torsional angle of ca. 80° out-of-plane [99]. In addition, other differences in hyperfine structure and spin density distribution relate to the degree of protonation at the phenolic oxygen, where e.g. the *E. coli* R2 is deprotonated, the mammalian R2 is hydrogen bonded and the *S. typhimurium* $R2_F$ is protonated [100, 101].

The details of the reaction have been studied during reconstitution of wild type R2 protein (cf. Fig. 10) and in the Y122F mutant protein by rapid freeze-quench EPR, Mössbauer, and ENDOR spectroscopies, as well as stopped-flow uv-vis and EPR spectroscopies [75, 96, 102–106a]. The following transient species have been observed:

i) A singlet EPR signal, with Mössbauer isomer shifts indicative of a spin-coupled Fe^{III}/Fe^{IV} species (denoted species X). Recent ENDOR studies have identified three oxygen ligands, most likely originating from the dioxygen molecule and solvent water. Species X may also be associated with a uv-vis absorption at 365 nm.

Prior to or concomitant with species X two other transient species have been observed:

ii) A transient uv-vis absorption at 410 nm observable in the Y122F protein but not in doubly mutant proteins Y122F/Y356A and Y122F/Δ30C (lacking residues 346–375), which has been ascribed to Tyr-356.
iii) A singlet EPR signal observable at room temperature, thereby distinguishing it from species X, and perhaps associated with a 560 nm uv-vis band, which has been tentatively assigned to Trp-48.

As both Tyr-356 and Trp-48 are believed to be members of a hydrogen bonded pathway in operation during catalysis, it has been suggested that this pathway may also be operating during reconstitution to transfer the fourth electron from solvent reductants. Surprisingly, the end products of the reconstitution reaction are also formed, albeit with about an order of magnitude lower rate constants, in mutant R2 proteins that affect the proposed electron transfer pathway (e.g. W48Y, D237N/E in *E. coli* R2 [M. Sahlin, unpublished; M. Ekberg, unpublished], and

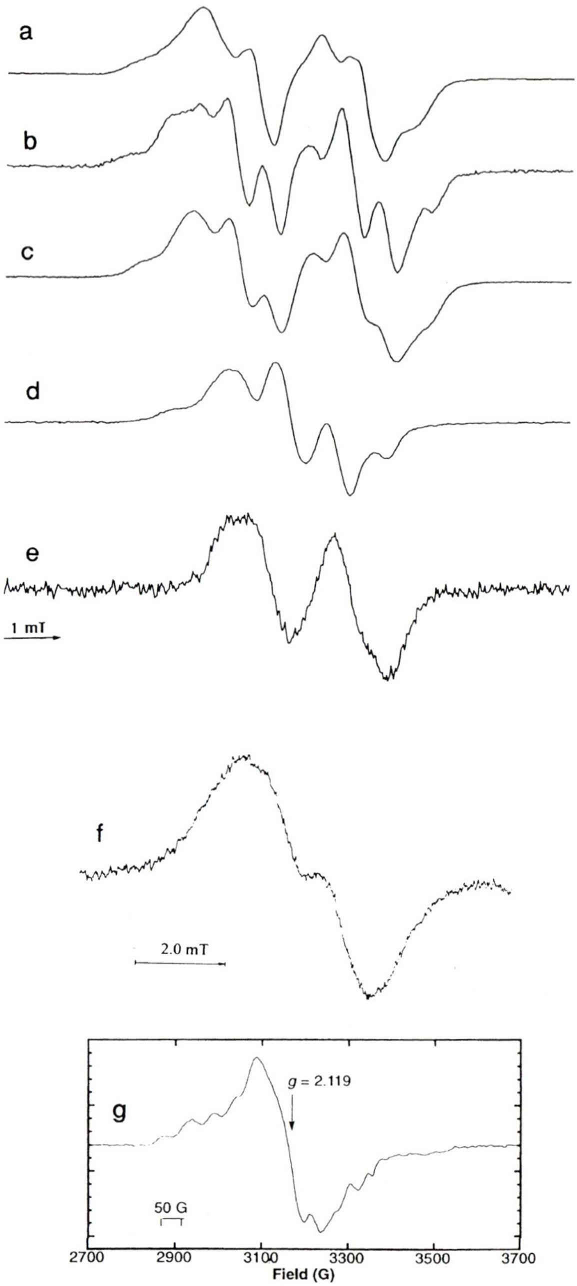

Fig. 9. EPR spectra of R2 tyrosyl radicals from **a** *E. coli,* **b** bacteriophage T4, **c** mouse class Ia reductase, **d** *S. typhimurium* class Ib reductase, **e** NrdD glycyl radical from bacteriophage T4 class III ribonucleotide reductase; **f** organic free radical of *C. ammoniagenes* ribonucleotide reductase, from [20]; **g** Co(II)-thiyl radical in *L. leichmannii*class II reductase, adapted from [124]

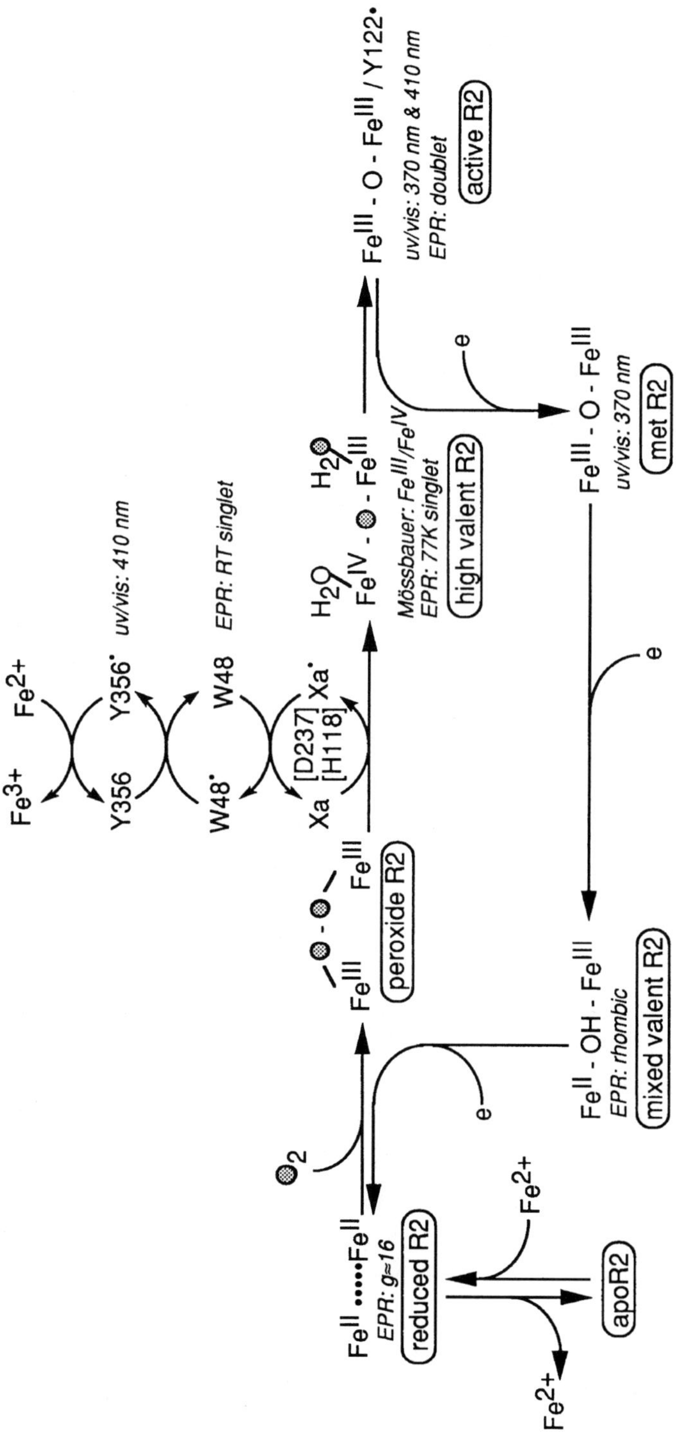

Fig. 10. Schematic dioxygen-dependent reconstitution mechanism, reduction cycle of diferrous protein R2

W103Y, D266A in mouse R2 [106b]). This implies that a fourth electron may be supplied by different means, but that the catalytically essential pathway is more efficient. In addition, mutations in *E. coli* R2 affecting some iron ligating side chains (D84A/H, E204A, E238A, and H241A) still yield a diiron site with absorption bands in the 320–370 nm region indicative of an oxo-bridged diferric site; the D84A, H118A, and E204A mutant proteins even form a transient tyrosyl radical, conceivably at Tyr-122 [45]. Thus, even the diiron site can accept seemingly drastic changes and still undergo similar chemical reaction sequences as occur in the reconstitution of wild-type R2 protein.

4.1.3
Other Chemical Reactions Performed by Mutationally Changed Diiron-oxo Sites

The diiron site of class I ribonucleotide reductase has structural and spectroscopic properties in common with several other diiron-oxo proteins [72, 73], e.g. the hydroxylase component of methane monooxygenase [107], bacterioferritin [108a, b], rubrerythrin [109] and fatty acid Δ9 desaturase [110]. The four helix bundle, wherein the iron binding motif – D/Exx*H* – occurs twice, is common to all these diiron proteins (Fig. 11). The structure determinations of these proteins include different redox states, e.g. diferrous, mixed-valent and diferric forms. These diiron-oxo proteins all catalyze iron-oxygen chemistry, but the end result is very different. In ribonucleotide reductase a stable tyrosyl radical is formed, in MMO a hydrocarbon is hydroxylated, in desaturase a double bond is inserted into a carboxylic acid and in ferritins iron is stored as a ferric-hydroxylate mineral core (the physiological function of rubrerythrin remains to be shown, but it has ferroxidase activity). One tantalizing question is: what chemical or structural features make the different diiron sites perform such different chemistry as result of oxygenation? Attempts are currently pursued, e.g. in the research groups of P. Nordlund and B.-M. Sjöberg at Stockholm University, to engineer one diiron site into another by genetic means.

Three different mutant proteins of *E. coli* R2 have proved to catalyze changed chemistry as compared to the wild-type enzyme (Fig. 12b–e):

i) The Y122F mutation, where a fraction of the high valent iron-oxo species that would in the wild-type protein generate the stable tyrosyl radical at Tyr-122 instead oxidizes a nearby tryptophan (Trp-111) and thereby opens up a pathway for sequential radical transfer involving yet another tryptophan side chain (Trp-107) [75, 102]. Recent studies with specific deuteration of side chains, ENDOR studies [111] and further site-directed mutagenesis have shown that a tyrosine side chain is more efficiently oxidized than a tryptophan, and that aromatic stacking may be as efficient as through-bond connections in this type of radical transfer reaction (Fig. 12c) [112].
ii) The most intriguing result was obtained with the F208Y construct [113, 114]. Phe-208 constitutes one component of the hydrophobic pocket enclosing the diiron site and the adjacent Tyr-122 side chain. It was with great sur-

Diiron-oxo binding motifs

```
                      79                                              123
Class I protein R2    YQTLLDSIQGRSPNVALLPLISI......PELETWVETWAFSETIHSRSYT
Methane monooxygenase FLEVGEYNAIAATGMLWDSAQAA.......EQKNGYLAQVLDEIRHTHQCA
Δ9 Desaturase         DMITEEALPTYQTMLNTLDGVRDETGASPTSWAIWTRAWTAEENRHGDLLN
Rubrerythrin          TAFAGESQARNRYNYFGGQAKKD....GFVQISDIFAETADQEREHAKRLF
Bacterioferritin      KLLGNELVAINQYFLHARMFKNW....GLKRLNDVEYHESIDEMKHADRYI
            Consensus -----e------------------------------------E--H---y-
                           *                                    *  *   ↑

                      199                                             246
Class I protein R2    SVNALEAIRFYVSFACSFAFAER   ELMEGNAKIIRLIARDEALHLTGTQ
Methane monooxygenase LQLVGEACFTNPLIVAVTEWAAA   NGDEITPTVFLSIETDELRHMANGY
Δ9 Desaturase         YTSFQERATFI SHGNTARQAKE   HGDIKLAQICGTIAADEKRHETAYT
Rubrerythrin          ASAAGEHHEYTEMYPSFARIARE   EGYEEIARVFASIAVAEEFHEKRFL
Bacterioferritin      SDLALELDGAK.NLREAIGYADS   VHDYVSRDMMIEILRDEEGHIDWLE
            Consensus -----E-----------------   ----------------E--H-----
                           *                                    *  *
```

Fig. 11. Amino acid sequence alignment of iron binding motifs in diiron-oxo proteins. Class I protein R2 from *E. coli*, methane monooxygenase from *Methylococcus capsulatus*, Δ9 destaurase from castor bean, rubrerythrin from *Desulphovibrio vulgaris*, bacterioferritin from *E. coli*. Only regions in the vicinity of the iron ligands are shown. An *arrow* denotes the site of the stable tyrosyl radical in R2,, an *asterisk* conserved iron ligands. The consensus line shows residues conserved in all aligned species (upper case), residues conserved only in some sequences (lower case)

prise that we purified a deep blue mutant protein, as opposed to the bright green colored wild-type protein, the color relating to the electronic absorption of its tyrosyl radical and diferric site. The color of the F208Y protein turned out to originate from a ferric catecholate adduct formed by an autocatalytic two electron oxidation of the Tyr-208 side chain to a dopa-208 (Fig. 12d). The chemistry of this mutant protein illustrates that very subtle changes in the vicinity of the diiron site can perturb the iron-oxo chemistry drastically.

iii) Preliminary structural and electronic absorption data show that other changes at the diiron site may give rise to similar type of chemistry. The double mutant protein Y122F/E238A, where both the wild type stable radical harboring side chain and an iron liganding side chain are deleted, also promotes oxidation of the side chain at position 208 [115]. Glu-238, as described above has a crucial role during formation of the diferric site as it is a bridging ligand in the diferrous form and a monodentate ligand of Fe2 in the diferric form. In this case the Phe-208 side chain is autocatalytically oxidized by the high valent iron-oxo intermediate in a two electron process that results in a *m*-OH-phenylalanine side chain, that ligates to the diiron site (Fig. 12e) and most likely gives rise to an absorption band at 475 nm observed in the modified protein. Again changes in the vicinity of the diiron site have proved to change drastically the chemistry of the highly reactive iron-oxo intermediate.

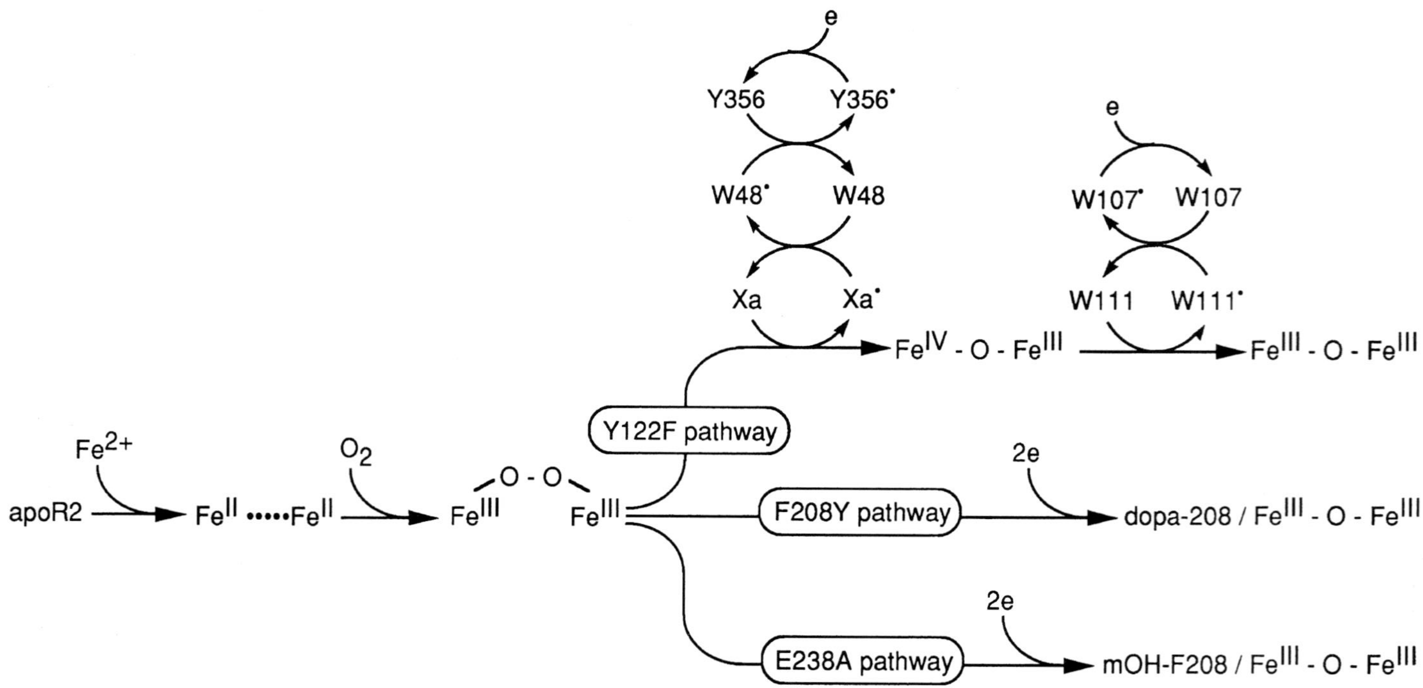

Fig. 12. **a** Proposed reaction scheme for iron reconstitution of some mutant R2 proteins. Three-dimensional structure of **b** diferric site in wild type protein R2, **c** hydrogen-bonded radical transfer pathways in W107Y/Y122F protein R2, **d** dopa adduct in F208Y protein R2, **e** *m*-OH-phenylalanine adduct in E238A/Y122F protein R2

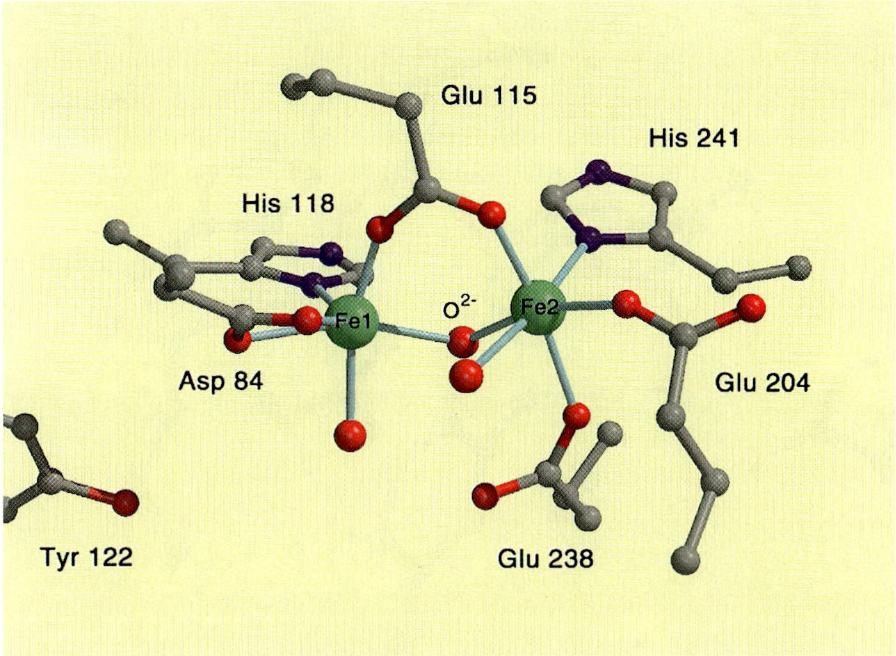

Fig. 12b

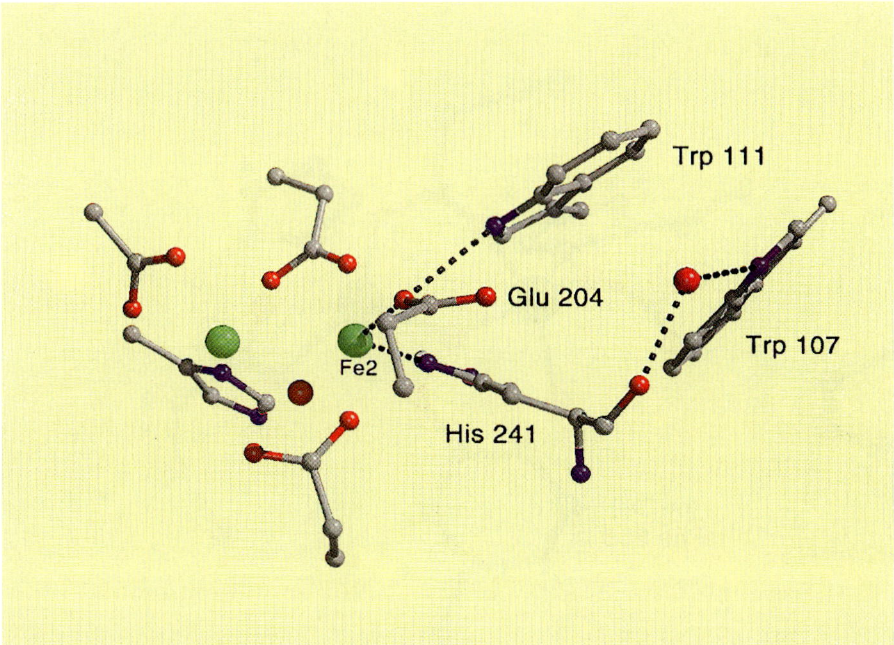

Fig. 12c

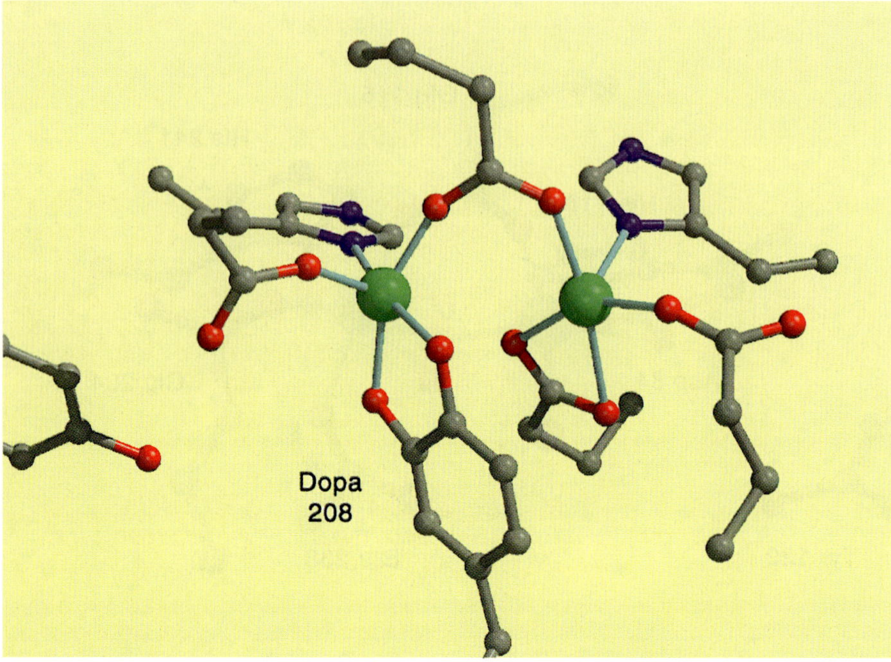

Fig. 12d

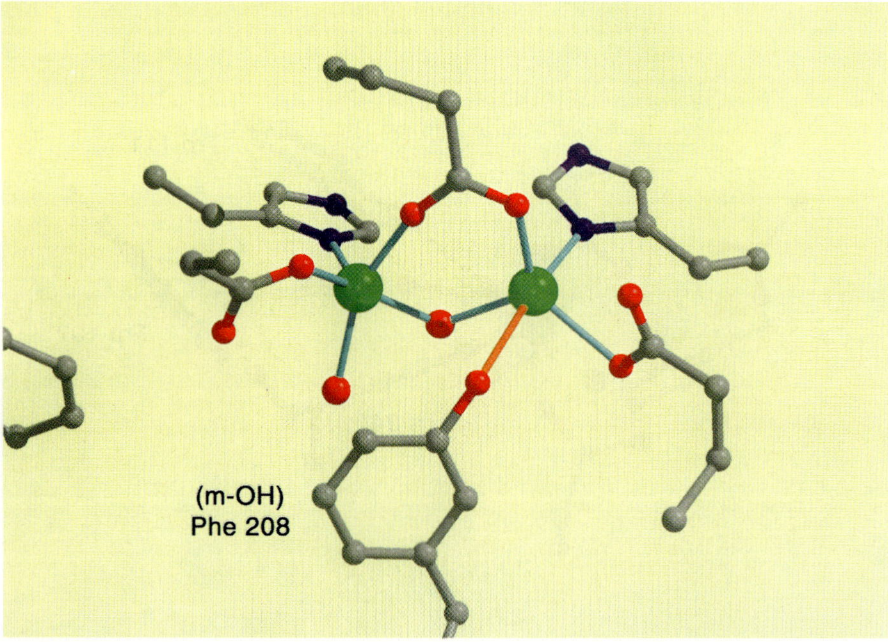

Fig. 12e

4.1.4
Binding of Other Metal Ions to the Diiron-oxo Site

Several other metal ions can bind to the empty dinuclear site in R2 proteins. None of these dinuclear complexes have had the capacity to generate a tyrosyl radical, implying that oxygen binding or oxidation of such a metal site is hampered.

The most extensively studied form is the dimanganese variant of *E. coli* R2, for which the three-dimensional structure was solved in 1992 [116]. The manganese ions were ligated to the same protein side chains as in the diiron form. The metal site in Mn-R2 is more symmetric than in diiron R2, with one monodentate carboxylate and one histidine (Asp-84/His-118 and Glu-204/His-241) per manganese ion, and Glu-115 and Glu-238 bridging the two metal ions. No other bridging ligand could be seen. This led to the proposal that Mn-R2 reflected the structure of the reduced diiron site in iron-containing R2, a hypothesis that the recent structure determination of reduced R2 has corroborate [83]. A striking difference between the dimanganese structure and the diferrous structure is the presence of a low molecular weight ligand (water or anionic salt) at Me2 in the former case; perhaps this external ligand perturbs the oxygen affinity of the former site as compared to the diiron form. Manganese ions have also been shown to bind to mouse R2, where this reaction has been used to compare metal binding strength between different mutant variants of mouse R2 [47].

A dinuclear site was also obtained by incubation of *E. coli* apoR2 with Co^{2+}. The resulting metallosite was shown spectroscopically to contain two five-coordinated Co^{II}, and was not oxidized in presence of air saturated buffers [117]. As for the dimanganese form, the two cobalt ions lack magnetic interaction, indicative of a structure with no solvent-derived oxo-bridge (or a protonated one). A particular beauty of the dinuclear cobalt cluster is that metal ligating side chains are visible in high resolution NMR spectroscopy, and a preliminary assignment has been performed [117].

4.2
Manganese Sites

It was recognized early on that some coryneform bacteria starved of manganese were inhibited in DNA synthesis and that addition of manganese to the starved cells resulted in elevated levels of ribonucleotide reductase activity, whereas addition of ferrous ions did not. It has also been shown that the *C. ammoniagenes* reductase harbors manganese ions and is inhibited by the radical scavenger hydroxyurea [10]. Recently, Griepenburg et al. [20] succeeded in characterizing the stable free radical of the *C. ammoniagenes* enzyme by EPR spectroscopy. The *g* value of this hyperfine singlet EPR signal is 2.0040 (Fig. 9), and its microwave saturation behavior indicates only very weak interactions with a metal site. The sample was prepared by extensive dialysis against EDTA to avoid interference from soluble manganese ions, and since the free radical was still present and observed to interact only weakly with a metal site, it is plausible that the presumed manganese site of *C. ammoniagenes* reductase is EPR silent, as would be expected for e.g. a coupled dinuclear site [10]. The presumed metal liganding residues of such a dinuclear manganese site are all conserved in the preliminary amino acid sequence deduced for the $R2_F$ protein of *C. ammoniagenes* [52]. A puzzling fact is that whereas manganese can bind and form a dinuclear site in e.g. class Ia enzymes, it does not promote generation of a free radical [47, 116]. Therefore, the binding of manganese in the *C. ammoniagenes* enzyme must clearly be different so as to allow radical generation. Studies with cloned and overproduced *C. ammoniagenes* reductase will probably be an efficient and enlightening way to understand the process.

4.3 Iron Sulfur Clusters

In 1993 the anaerobic ribonucleotide reductase of *E. coli* was shown to contain stoichiometric amounts of iron and sulfide, indicative of an iron-sulfur center [118]. It was not, however, until the two proteins constituting the NrdD:NrdG holoenzyme could be expressed and purified separately [58, 119] that it was shown that an Fe_4S_4 cluster is bound by the homodimeric NrdG component [19]. The separately purified NrdD component does not contain any iron-sulfur cluster. However, as previous data showed the presence of a Fe_3S_4 cluster in the NrdD:NrdG complex [118], it is still an open question as to whether the iron-sulfur cluster also remains bound to the NrdG component in the holoenzyme complex, or if some exchange of iron ligands occur during holoenzyme formation.

The Fe_4S_4 cluster characterized in NrdG has a prominent electronic absorption band at 420 nm with molar extinction coefficient of ca. 3500 $M^{-1}\,cm^{-1}$ and two less prominent bands at 460 nm and 550 nm (Fig. 13a). Its rhombic EPR signal has features at $g = 2.02, 1.92$, and 1.88 (Fig. 13b), and a microwave power saturation and temperature dependence typical of an Fe_4S_4 center.

As the NrdG component only forms an iron-sulfur cluster in its dimeric form whereas the monomeric form binds Fe_2S_2, it has been suggested that ligands to the FeS-cluster are contributed by both polypeptides [19, S. Ollagnier, personal communication]. In the prokaryotic anaerobic regulatory protein FNR, known to undergo monomeric-dimeric transition, Lazazzera et al. recently showed that the dimeric form binds one Fe_4S_4 cluster per monomer, and that oxygen inactivation of the highly labile iron-sulfur cluster promotes monomer formation [120]. Comparison of five known NrdG polypeptide sequences reveals three conserved cysteine residues (Fig. 6). In preliminary experiments all of these were shown to be needed for radical generation in the phage NrdD component [J. Andersson, unpublished]. More detailed studies are definitely required before any firm conclusions can be drawn about the structure, function, and dynamics of the iron-sulfur cluster in class III ribonucleotide reductase.

The functionally and structurally related Pfl activase also binds ferrous ions, and recently a Fe_4S_4 species has been spectroscopically characterized [121, 122[a,b]]. Based on the presence of several conserved cysteine residues in the *N*-terminal region of the Pfl activase (Fig. 6) it has been suggested that these side chains are the ligands of the metal center. In light of the striking similarity between the iron-binding NrdG, Pfl activase and FNR proteins regarding oxygen-sensitivity, monomer/dimer transition and some of the suggested metal

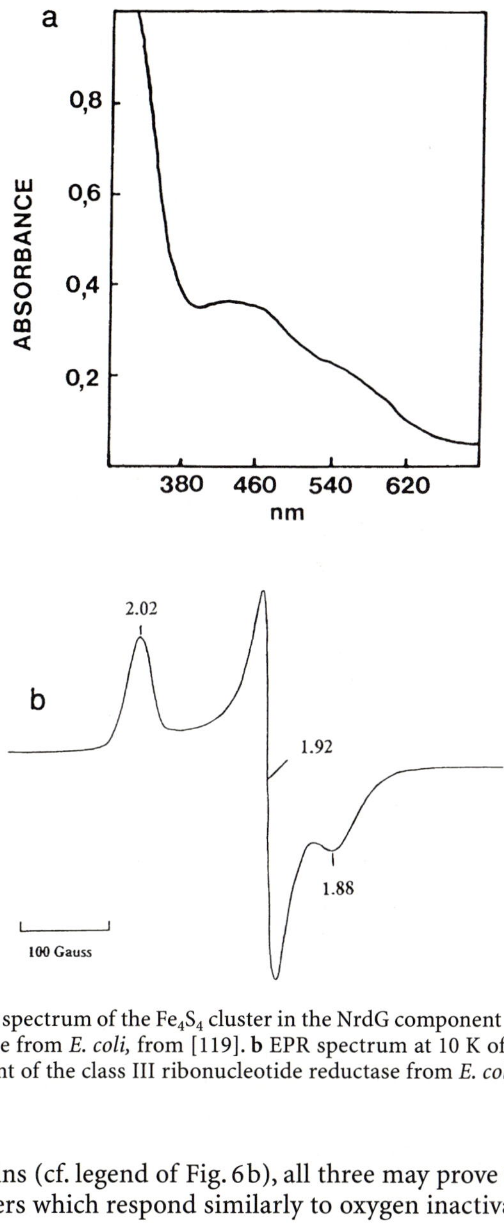

Fig. 13. **a** Electronic spectrum of the Fe_4S_4 cluster in the NrdG component of the class III ribonucleotide reductase from *E. coli*, from [119]. **b** EPR spectrum at 10 K of the Fe_4S_4 cluster in the NrdG component of the class III ribonucleotide reductase from *E. coli*, from [19]

ligating side chains (cf. legend of Fig. 6b), all three may prove to have common iron-sulfur clusters which respond similarly to oxygen inactivation.

4.4 Adenosyl Cobalamin Sites

A transient EPR spectrum was characterized in ribonucleotide reductase from *L. leichmannii* already in 1974 [123], and ascribed to a radical interacting with cob(II)alamin. Recently, Stubbe et al. [21, 124] have used enzyme enriched in

β-deuterated cysteine coupled with spectral simulations to show that the EPR species arises from interaction between a cysteinyl radical and the cob(II)alamin at a distance of ca. 5–8 Å. This distance allows for a direct interaction between the cysteine side chain, presumably Cys-408 and the deoxyadenosyl part of the cofactor, which presumably forms a transient radical too short-lived to be detected. However, the magnetic interactions observed are also compatible with a less direct interaction, e.g. a short highly conjugated network, between the transient cysteinyl radical at the active site and the Co^{II} site.

Recent studies on B_{12}-dependent enzymes have revealed that both an enzyme catalyzing heterolytic cleavage (methyl transfer reaction) and an enzyme catalyzing homolytic cleavage at the upper axial position of the corrin ring have similar arrangements of side chains in the domain that interacts with the lower face of the corrin ring: a His-Asp-Ser triad in the methyl transferase and a His-Asp-Lys triad in the mutase [125]. The structure-based sequence motif -**Asp**-X-**His**-X-X-Gly–(41)–**Ser**-X-Leu–(26–28)–Gly-Gly- (triad components in **bold**) was identified in several of the B_{12}-dependent enzymes [125, 126]. The critical histidine residue displaces the dimethylbenzimidazole ligand when the cofactor is bound, and it is believed that the chemical properties of the histidine-cobalt bond, as manifested by the different triads, determines whether hetero- or homolytic cleavage occurs.

In class II enzyme from *L. Leichmannii* a homolytic cleavage occurs. It is however not possible to find the B_{12}-binding fingerprint in a sequence comparison between the class II enzymes. A remnant may be the conserved region -**Asp**-X-**His**–(14–16)–**Ser**-X-Ile–(10)–Gly-Gly- occurring between positions 226–259 in the *L. leichmannii* sequence. In the archeal sequences there are only two conserved histidines; one occurs in the overall consensus -**Ser**-X-Leu–(33)–Gly-Gly–(13)–**His**-X-**Asp**-, again distantly reminiscent of a B_{12}-binding fingerprint.

5
Conclusions and Perspectives

The reaction catalyzed by ribonucleotide reductases, i.e. the production of building blocks for storage of genetic information, is beyond doubt one of the earliest essential reactions to life on Earth. Although seemingly diverse, this group of enzymes builds on the unifying theme of radical chemistry. As recently demonstrated by the isolation and sequencing of two archaebacterial ribonucleotide reductases, the substrate binding component of the enzyme conceivably has a common evolutionary origin. In contrast, the radical generating entities are diverse (e.g. cobalamin cofactor, tyrosyl radical, glycyl-radical) and may during evolution have been picked up from available entities with similar capacities: cobalamin is also known to undergo homolytic cleavage to generate a deoxyadenosyl radical in B_{12}-dependent mutases [127], adoMet is, in the Pfl system, known to generate a deoxyadenosyl radical by homolysis [67], diiron-oxo sites are known to involve radical reactions in methane monooxygenase [128]. It is therefore conceivable that the contemporary radical generating mechanisms of ribonucleotide reductases evolved later than the enzyme proper.

One might speculate that ribonucleotide reductases in which the unpaired electron needed for catalysis is generated in the actual substrate binding component (class II and III) are closer to an ur-reductase than those that stabilize the unpaired electron in another component (class I). In addition, the phosphorylation level of substrates and allosteric effectors may have been the same in a less sophisticated ribonucleotide reductase, and metabolically closer to the products needed for DNA synthesis. Some non-archeal representatives of class II and the class III representatives are more ancient also in this respect, as they reduce ribonucleoside triphosphates and the corresponding deoxynucleotide products act as allosteric effectors. If a class II respresentative were the most primordial enzyme, one may wonder why Nature ever bothered to develop other forms. Of course, dearth of AdoCbl would promote other forms, but a more convincing cause is perhaps that the class II enzyme has been shown to generate a highly reactive active site cysteinyl radical even in the absence of substrate. Such a reaction would both cause unnecessary energy waste and potential radical mediated degradation of the reductase proper. Thus, a separation in time and space of the radical generating process and the reduction of ribonucleotides as in class I and III seems conceptually superior. It is of interest that both class II and III representatives are found in *Archea.*

Another exciting finding is that some of the archaebacterial ribonucleotide reductase gene sequences encode inteins (i.e. coding for self-splicing protein sequences) [13, 14], whereas some bacteriophage ribonucleotide reductase gene sequences encode group I introns (i.e. coding for self-splicing pre-mRNA sequences) [59, 129]. It still remains an enigma whether prokaryotic occurrences of group I introns are the result of vertical or horizontal inheritance, but their occurrence in bacteriophages are restricted to enzymes involved in DNA synthesis. An observation that may have far reaching consequences in an evolutionary perspective is that the insertion site of one intein in *M. jannaschii* class III ribonucleotide reductase differs by only 11 residues from the insertion site of the intron in bacteriophage T4 NrdD component. This may imply that group I introns and inteins have a common evolutionary origin and perhaps are molecular fossils of a primordial gene regulatory mechanism.

Acknowledgements. I thank D. Logan and M. Sahlin for invaluable help with illustrations, A. Gräslund, M. Karlsson, D. Logan, P. Nordlund and M. Sahlin for constructive criticism, and D. Logan for excellent linguistic help. Cited work performed in my own research group was supported by grants from the Swedish Cancer Society, the Swedish Research Councils for Natural Sciences and Engineering Sciences, and the Swedish National Board for Technical and Industrial Development.

6 References

1. Lin AL, Elford HL (1980) J Biol Chem 255:8523–8528
2. Markert ML (1991) Immunodefic Rev 3:45–81
3. Caras IW, Martin DWJ (1988) Mol Cell Biol 8:2698–2704
4. Nocentini G (1996) Crit Rev Oncol Hematol 22:89–126
5. Reichard P (1989) Biochim Biophys Acta 1000:49–50
6. Reichard P (1995) Annu Rev Biochem 64:1–28
7. Blakley RL, Orme-JohnsonWH, Bozdech JM (1979) Biochemistry 18:2335–2339
8. Gleason FK, Hogenkamp HP (1972) Biochim Biophys Acta 277:466–470
9. Schimpff WG, Follmann H, Auling G (1981) Biochem Biophys Res Commun 102: 1276–1282
10. Willing A, Follmann H, Auling G (1988) Eur J Biochem 170:603–611
11. Fontecave M, Eliasson R, Reichard P (1989) Proc Natl Acad Sci USA 86:2147–2151
12. Sun XY, Ollagnier S, Schmidt PP, Atta M, Mulliez E, Lepape L, Eliasson R, Gräslund A, Fontecave M, Reichard P, Sjöberg BM (1996) J Biol Chem 271:6827–6831
13. Bult CJ, White O, Olsen GJ, Zhou L, Fleischmann RD, Sutton GG, Blake JA, FitzGerald LM, Clayton RA, Gocayne JD, Kerlavage AR, Dougherty BA, Tomb JF, Adams MD, Reich CI, Overbeek R, Kirkness EF, Weinstock KG, Merrick JM, Glodek A, Scott JL, Geoghagen NSM, Weidman JF, Fuhrmann JL, Venter JC et al. (1996) Science 273:1058–1073
14. Riera J, Robb FT, Weiss R, Fontecave M (1996) Proc Natl Acad Sci USA 94:475–478
15. Tauer A, Benner SA (1996) Proc Natl Acad Sci USA 94:53–58
16. Petersson L, Gräslund A, Ehrenberg A, Sjöberg B-M, Reichard P (1980) J Biol Chem 255: 6706–6712
17. Nordlund P, Sjöberg B-M, Eklund H (1990) Nature 345:593–598
18. Nordlund P, Eklund H (1993) J Mol Biol 232:123–164
19. Ollagnier S, Mulliez E, Gaillard J, Eliasson R, Fontecave M, Reichard P (1996) J Biol Chem 271:9410–9416
20. Griepenburg U, Lassmann G, Auling G (1996) Free Radical Res 24:473–481
21. Gerfen GJ, Licht S, Willems JP, Hoffman BM, Stubbe J (1996) J Am Chem Soc 118: 8192–8197
22. Åberg A, Hahne S, Karlsson M, Larsson A, OrmöM, Åhgren A, Sjöberg B-M (1989) J Biol Chem 264:12249–12252
23. Mao SS, Yu GX, Chalfoun D, Stubbe J (1992) Biochemistry 31:9752–9759
24. Stubbe J, Ator M, Krenitsky T (1983) J Biol Chem 258:1625–1631
25. Mao SS, Holler TP, Yu GX, Bollinger JM, Booker S, Johnston MI, Stubbe J (1992) Biochemistry 31:9733–9743
26. Ashley GW, Harris G, Stubbe J (1986) J Biol Chem 261:3958–3964
27. Sjöberg B-M, Gräslund A, Eckstein F (1983) J Biol Chem 258:8060–8067
28. Behravan G, Sen S, Rova U, Thelander L, Eckstein F, Gräslund A (1995) BBA-Gene Struct Express 1264:323–329
29. van der Donk WA, Stubbe J, Gerfen GJ, Bellew BF, Griffin RG (1995) J Am Chem Soc 117: 8908–8916
30. Covés J, Defallois LLH, Lepape L, Decout JL, Fontecave M (1996) Biochemistry 35: 8595–8602
31. van der Donk WA, Zeng CH, Biemann K, Stubbe J, Hanlon A, Kyte J (1996) Biochemistry 35:10058–10067
32. Holmgren A, Björnstedt, M (1995) In: Packer L (ed) Biothiols, Vol 252, pp 199–208 Academic Press Inc, San Diego, CA 92101–4495
33. Holmgren A, Åslund F (1995) In: Packer L (ed) Biothiols, Vol 252, pp 283–292 Academic Press Inc, San Diego, CA 92101–4495
34. Uhlin U, Eklund H (1994) Nature 370:533–539
35. Reichard P (1993) Science 260:1773–1777

36. Sjöberg B-M (1995) In: Eckstein F, Lilley DMJ (eds) Nucl Acids Mol Biol, Vol 9, pp 192–221 Springer, Berlin Heidelberg New york
37. Uhlin U, Eklund H (1996) J Mol Biol 262:358–369
38a. Ormö M, Sjöberg B-M (1996) Eur J Biochem 241:363–367
38b. Eriksson M, Uhlin U, Ramaswamy S, Ekberg M, Regnström K, Sjöberg B-M, Eklund H (1997) Structure
39. Nielsen BB, Kauppi B, Thelander M, Thelander L, Kjøller Larsen I, Eklund H (1995) FEBS Lett 373:310–312
40. Kauppi B, Nielsen BA, Ramaswamy S, Kjøller Larsen I, Thelander M, Thelander L, Eklund H (1996) J Mol Biol 262:706–720
41. Ehrenberg A, Reichard P (1972) J Biol Chem 247:3485–3488
42. Larsson A, Sjöberg BM (1986) EMBO J 5:2037–2040
43. Sjöberg BM (1994) Structure 2:793–796
44. Climent I, Sjöberg B-M, Huang CY (1992) Biochemistry 31:4801–4807
45. Persson BO, Karlsson M, Climent I, Ling J, Sanders-Loehr J, Sahlin M, Sjöberg B-M (1996) J Biol Inorg Chem 1:247–256
46. Ekberg M, Sahlin M, Eriksson M, Sjöberg BM (1996) J Biol Chem 271:20655–20659
47. Rova U, Goodtzova K, Ingemarson R, Behravan G, Gräslund A, Thelander L (1995) Biochemistry 34:4267–4275
48. Jordan A, Pontis E, Atta M, Krook M, Gibert I, Barbé J Reichard P (1994) Proc Natl Acad Sci USA 91:12892–12896
49. Jordan A, Gibert I, Barbé J (1995) Gene 167:75–79
50. Jordan A, Pontis E, Åslund F, Hellman U, Gibert I, Reichard P (1996) J Biol Chem 271: 8779–8785
51. Fraser CM, Gocayne JD, White O, Adams MD, Clayton RA, Fleischmann RD, Bult CJ, Kerlavage AR, Sutton G, Kelley JM, Fritchman JL, Weidman JF, Small KV, Sandusky M, Fuhrmann J, Nguyen D, Utterback TR, Saudek DM, Phillips CA, Merrick JM, Tomb JF, Dougherty BA, Bott KF, Hu PC, Lucier TS, Peterson SN, Smith HO, Hutchison CA, Venter JC (1995) Science 270:397–403
52. Tulokhonova L, Torrents Serra E, Jordan A, Gibert I, Hellman U, Barbé J, Karlsson M, Sjöberg B-M (1997) Manuscript:
53. Booker, S Stubbe, J (1993) Proc Natl Acad Sci U S A 90:8352–8356
54. Hatfull GF, Sarkis GJ (1993) Mol Microbiol 7:395–405
55. Hogenkamp HPC, McFarlan SC (1989) In: Cory JG, Cory AH, (eds) Inhibitors of ribonucleoside diphosphate reductase activity, pp 17–36 Pergamon Press, Inc, Exeter, UK
56. Kaneko T, Sato S, Kotani H, Tanaka A, Asamizu E, Nakamura Y, Miyajima N, Hirosawa M, Sugiura M, Sasamoto S, Kimura T, Hosouchi T, Matsuno A, Muraki A, Nakazaki N, Naruo K, Okumura S, Shimpo S, Takeuchi C, Wada T, Watanabe A, Yamada M, Yasuda M, Tabata S (1996) DNA Res 3:109–136
57. Booker S, Licht S, Broderick J, Stubbe J (1994) Biochemistry 33:12676–12685
58. Sun X, Harder J, Krook M, Jörnvall H, Sjöberg BM, Reichard P (1993) Proc Natl Acad Sci USA 90:577–581
59. Young P, Öhman M, Xu MQ, Shub DA, Sjöberg BM (1994) J Biol Chem 269:20229–20232
60. Young P, Öhman M, Sjöberg BM (1994) J Biol Chem 269:27815–27818
61. Young P, Andersson J, Sahlin M, Sjöberg BM (1996) J Biol Chem 271:20770–20775
62. Eliasson R, Pontis E, Sun XY, Reichard P (1994) J Biol Chem 269:26052–26057
63. Harder J, Eliasson R, Pontis E, Ballinger MD, Reichard P (1992) J Biol Chem 267: 25548–25552
64. Bianchi V, Reichard P, Eliasson R, Pontis E, Krook M, Jörnvall H, Haggård-Ljungquist E (1993) J Bacteriol 175:1590–1595
65. Mulliez E, Ollagnier S, Fontecave M, Eliasson R, Reichard P (1995) Proc Natl Acad Sci USA 92:8759–8762
66. Wagner AF, Frey M, Neugebauer FA, Schäfer W, Knappe J (1992) Proc Natl Acad Sci USA 89:996–1000
67. Frey M, Rothe M, Wagner AFV, Knappe J (1994) J Biol Chem 269:12432–12437

68. Knappe J, Wagner AFV (1995) In: Klinman JP (ed) Redox - Active Amino Acids in Biology, Vol 258, pp 343-362 Academic Press, San Diego, CA 92101-4495
69. Parast CV, Wong KK, Kozarich JW, Peisach J, Magliozzo RS (1995) Biochemistry 34: 5712-5717
70. Parast CV, Wong KK, Kozarich JW, Peisach J, Magliozzo RS (1995) J Am Chem Soc 117: 10601-10602
71. Atkin CL, Thelander L, Reichard P, Lang G (1973) J Biol Chem 248:7464-7472
72. Nordlund P, Eklund H (1995) Curr Opin Struct Biol 5:758-766
73. Andersson KK, Gräslund A (1995) Adv Inorg Chem 43:359-407
74. Silva KE, Elgren TE, Que L, Stankovich MT (1995) Biochemistry 34:14093-14103
75. Sahlin M, Lassmann G, Pötsch S, Sjöberg BM, Gräslund A (1995) J Biol Chem 270: 12361-12372
76. Tong WH, Chen S, Lloyd SG, Edmondson DE, Huynh BH, Stubbe J (1996) J Am Chem Soc 118:2107-2108
77. Backes G, Sahlin M, Sjöberg BM, Loehr TM, Sanders-Loehr J (1989) Biochemistry 28: 1923-1929
78. Ormö M, Regnström K, Wang ZG, Que L, Sahlin M, Sjöberg BM (1995) J Biol Chem 270: 6570-6576
79. Kjøller Larsen I, Sjöberg BM, Thelander L (1982) Eur J Biochem 125:75-81
80. Pötsch S, Drechsler H, Liermann B, Gräslund A, Lassmann G (1994) Mol Pharmacol 45: 792-796
81. Pötsch S, Sahlin M, Langelier Y, Gräslund A, Lassmann G (1995) FEBS Lett 374:95-99
82. Åberg A, Nordlund P, Eklund H (1993) Nature 361:276-278
83. Logan DT, Su XD, Åberg A, Regnström K, Hajdu J, Eklund H, Nordlund, P (1996) Structure 4:1053-1064
84. Sahlin M, Gräslund A, Petersson L, Ehrenberg A, Sjöberg BM (1989) Biochemistry 28: 2618-2625
85. Covés J, Fontecave M (1993) Eur J Biochem 211:635-641
86. Fontecave M, Eliasson R, Reichard P (1987) J Biol Chem 262:12325-12331
87. Su X-D, Persson B-O, Sjöberg B-M, Nordlund P (1996) Biochemistry Manuscript:
88. Atta M, Andersson KK, Ingemarson R, Thelander L, Gräslund A (1994) J Am Chem Soc 116:6429-6430
89. Gerez C, Gaillard J, Latour J-M, Fontecave M (1991) Angew Chem Int Ed Eng 30: 1135-1136
90. Gerez C, Fontecave M (1992) Biochemistry 31:780-786
91. Hendrich MP, Elgren TE, Que LJ (1991) Biochem Biophys Res Commun 176:705-710
92. Davydov R, Kuprin S, Gräslund A, Ehrenberg A (1994) J Am Chem Soc 116:11120-11128
93. Davydov R, Sahlin M, Kuprin S, Gräslund A, Ehrenberg A (1996) Biochemistry 35: 5571-5576
94. Davydov RM, Ménage S, Fontecave M, Gräslund A, Ehrenberg A (1997) Inorg Chem in press
95. Ling JS, Sahlin M, Sjöberg B-M, Loehr TM, Sanders-Loehr J (1994) J Biol Chem 269: 5595-5601
96. Bollinger JM, Tong WH, Ravi N, Huynh BH, Edmondson DE, Stubbe J (1994) J Am Chem Soc 116:8015-8023
97. Bender CJ, Sahlin M, Babcock GT, Barry BA, Chandrashekar TK, Salowe SP, Stubbe J, Lindström B, Petersson L, Ehrenberg A, Sjöberg BM (1989) J Am Chem Soc 111: 8076-8083
98. Hoganson CW, Sahlin M, Sjöberg BM, Babcock GT (1996) J Am Chem Soc 118: 4672-4679
99. Himo F, Gräslund A, Eriksson LA (1997) Biophys J 72:1556-1567
100. Schmidt PP, Andersson KK, Barra AL, Thelander L, Gräslund A (1996) J Biol Chem 271: 23615-23618
101. Allard P, Barra AL, Andersson KK, Schmidt PP, Atta M, Gräslund A (1996) J Am Chem Soc 118:895-896

102. Sahlin M, Lassmann G, Pötsch S, Slaby A, Sjöberg B-M, Gräslund A (1994) J Biol Chem 269:11699–11702
103. Ravi N, Bollinger JM, Huynh BH, Edmondson DE, Stubbe J (1994) J Am Chem Soc 116: 8007–8014
104. Bollinger JM, Tong WH, Ravi N, Huynh BH, Edmondson DE, Stubbe J (1994) J Am Chem Soc 116:8024–8032
105. Burdi D, Sturgeon BE, Tong WH, Stubbe JA, Hoffman BM (1996) J Am Chem Soc 118: 281–282
106a. Sturgeon BE, Burdi D, Chen SX, Huynh BH, Edmondson DE, Stubbe J, Hoffman BM (1996) J Am Chem Soc 118:7551–7557
106b. Schmidt PP, Rova U, Katterle B, Thelander L, Gräslund A (1997) J Biol Chem
107. Rosenzweig AC, Frederick CA, Lippard SJ, Nordlund P (1993) Nature 366:537–543
108a. Frolow F, Kalb AJ, Yariv J (1994) Nat Struct Biol 1:453–460
108b. LeBrun N, Wilson MT, Andrews SC, Guest JR, Harrison PM, Thomson AJ, Moore GR (1993) FEBS Lett 333:197–202
109. deMaré F, Kurtz DM, Nordlund P (1996) Nature Struct Biology 3:539–546
110. Lindqvist Y, Huang WJ, Schneider G, Shanklin J (1996) EMBO J 15:4081–4092
111. Lendzian F, Sahlin M, Macmillan F, Bittl R, Fiege R, Pötsch S, Sjöberg BM, Gräslund A, Lubitz W, Lassmann G (1996) J Am Chem Soc 118:8111–8120
112. Katterle B, Sahlin M, Schmidt PP, Pötsch S, Logan D, Gräslund A, Sjöberg B-M (1997) J Biol Chem 272:10414–10421
113. Ormö M, deMaré F, Regnström K, Åberg A, Sahlin M, Ling J, Loehr TM, Sanders-Loehr J, Sjöberg BM (1992) J Biol Chem 267:8711–8714
114. Åberg A, Ormö M, Nordlund P, Sjöberg BM (1993) Biochemistry 32:9845–9850
115. Logan D, Persson BO, Sjöberg B-M, Nordlund P (1997) Manuscript:
116. Atta M, Nordlund P, Åberg A, Eklund H, Fontecave M (1992) J Biol Chem 267: 20682–20688
117. Elgren TE, Ming LJ, Que LJ (1994) J Am Chem Soc 33:891–894
118. Mulliez E, Fontecave M, Gaillard J, Reichard P (1993) J Biol Chem 268:2296–2299
119. Sun XY, Eliasson R, Pontis E, Andersson J, Buist G, Sjöberg BM, Reichard P (1995) J Biol Chem 270:2443–2446
120. Lazazzera BA, Beinert H, Khoroshilova N, Kennedy MC, Kiley PJ (1996) J Biol Chem 271: 2762–2768
121. Wong KK, Murray BW, Lewisch SA, Baxter MK, Ridky TW, Ulissi-DeMario L, Kozarich JW (1993) Biochemistry 32:14102–14110
122a. Knappe J, Sawers G (1990) Fems Microbiol Rev 6:383–398
122b. Broderick JB, Duderstadt RE, Fernandez DC, Wojtuszewski K, Henshaw TF, Johnson MK (1997) J Am Chem Soc, in press
123. Orme-Johnson WH, Beinert H, Blakley RL (1974) J Biol Chem 249:2338–2343
124. Licht S, Gerfen GJ, Stubbe JA (1996) Science 271:477–481
125. Ludwig ML, Drennan CL, Matthews RG (1996) Structure 4:505–512
126. Drennan CL, Huang S, Drummond JT, Matthews RG, Ludwig ML (1994) Science 266: 1669–674
127. Mancia F, Evans PR (1996) Structure 4:339–350
128. Nesheim JC, Lipscomb JD (1996) Biochemistry 35: 0240–10247
129. Sjöberg BM, Hahne S, Mathews CZ, Mathews CK, Rand KN, Gait MJ (1986) EMBO J 5: 2031–2036

Protein Engineering of Cytochrome $P450_{cam}$

Luet-Lok Wong,* Andrew C.G. Westlake, and Darren P. Nickerson

Inorganic Chemistry Laboratory, South Parks Road, Oxford OX1 3QR, U.K.
**E-mail: luet.wong@icl.ox.ac.uk*

The cytochrome P450 superfamily of heme monooxygenases catalyse many reactions involved in the biosynthesis and degradation of endogenous compounds and in the oxidative metabolism of xenobiotics. These enzymes all share the same overall catalytic mechanism, and their varied substrate specificities arise from differences in their active site topologies, suggesting that it should be possible to engineer P450 enzymes for the oxidation of unnatural substrates. A review of the active site protein engineering experiments on cytochrome $P450_{cam}$ reported to date is presented here. Cytochrome $P450_{cam}$ is a stable, soluble enzyme isolated from *Pseudomonas putida*, and is an excellent system for rational protein engineering. The current understanding of the details of the $P450_{cam}$ catalytic mechanism, including the uncoupling pathways, is first discussed. The factors controlling substrate access, binding and turnover activity, and the coupling efficiency have been investigated by site-directed mutagenesis. The results of this work have enabled the rational redesign of $P450_{cam}$ to broaden the substrate range, to improve the activity and coupling efficiency towards unnatural substrates, and to engineer the regioselectivity of substrate oxidation. The construction of a "self-sufficient" fusion protein containing all three proteins of the $P450_{cam}$ system brings one step nearer the in vitro and in vivo biotechnological applications of $P450_{cam}$.

Keywords: Heme monooxygenases; cytochrome $P450_{cam;}$ unnatural substrates; mutagenesis; protein engineering.

1 Introduction

1.1 The Catalytic Activity of Cytochrome P450 Enzymes

Cytochrome P450 enzymes are found in almost all living organisms, including bacteria [1–3], yeast [4], plants [5, 6] and mammals [7]. They catalyse primarily the oxidation of unactivated carbon-hydrogen bonds to an alcohol functionality [8, 9], one of the most difficult reactions to achieve using conventional methods of chemical synthesis. In addition, P450 enzymes also catalyse the transfer of an oxygen atom to heteroatoms, the epoxidation of olefins and aromatic hydrocarbons, and the deep oxidation of alkyl groups leading ultimately to dealkylation. The substrate range of P450 enzymes is extremely broad, ranging from simple alkanes to the most complex of hormones and their precursors.

Many oxidation reactions catalysed by P450 enzymes are crucial to the growth, survival and maintenance of the organism. For instance, certain bacteria [10, 11] and strains of yeast [12] can grow on chemically inert substances such as alkanes by using P450 enzymes to oxidize them to more reactive intermediates as the first step in the catabolic pathways. P450 enzymes are involved in the biosynthesis and biodegradation of many endogenous compounds such as fatty acids, steroids and prostaglandins [13, 14], and they also play central roles in the oxidative degradation of chemically inert xenobiotics such as drugs and environmental contaminants [15]. The oxidation of xenobiotics converts them to more soluble and reactive derivatives which may be more easily eliminated from the organism, and some drugs are actually converted to their pharmacologically active forms by the action of P450 enzymes. However, some procarcinogens are activated by P450-mediated oxidation, e.g. benzo[*a*]pyrene and other polycyclic aromatic hydrocarbons (via diol-epoxide formation) [16, 17] and aryl and heterocyclic amines (via *N*-hydroxylation) [18].

The catalytic activity of P450 enzymes, which occurs under ambient conditions and often with high substrate specificity and regio- and stereo-selectivity, has captured the imagination of chemists and biochemists alike. The ability to harness this reactivity could have significant applications in the chemical and pharmaceutical industries, and in environmental bioremediation.

1.2
Eukaryotic and Prokaryotic P450 Enzymes

Eukaryotic P450 enzyme are normally membrane-associated systems. Because of the inherent difficulties in purifying and reconstituting often small quantities of such systems in vitro, much of our current knowledge of the catalytic mechanism and structure-function relationships of P450 enzymes is derived from work on soluble bacterial systems and one of the most amenable of the eukaryotic enzymes, rat liver $P450_{LM2}$. With recent rapid advances in the methodology of protein expression, many eukaryotic P450 enzymes have now been expressed in mammalian cell lines [19], baculovirus [20], yeast [21], and *E. coli* [22], and the techniques of reconstituting the catalytic activity of these membrane-associated recombinant forms are also being refined to the extent that significant advances can be expected in the near future [23, 24].

Prokaryotic P450 enzymes are involved in biosynthesis and biodegradation reactions as well as in catabolic pathways. Soluble P450 enzymes are found in some prokaryotes, and because of their relative ease of isolation from the organism, heterologous expression in *E. coli*, and crystallization, these have been studied in detail. The camphor hydroxylase cytochrome $P450_{cam}$ from *Pseudomonas putida* [25, 26] has been the model system from which much of our knowledge and current concepts of the mechanism of P450 catalysis are derived [27], and was the first P450 for which the X-ray crystal structure was determined. Other structurally characterised bacterial P450 enzymes include $P450_{terp}$ which enables a *Pseudomonad* to grow on α-terpinol [28], $P450_{BM\text{-}3}$ from *Bacillus megaterium* which is a fatty acid hydroxylase [29], and $P450_{eryF}$ from *Saccharopolyspora erythaea* which is involved in the biosynthesis of the antibiotic erythromycin A [30].

All P450 enzymes have an iron-protoporphyrin IX prosthetic group with a cysteine thiolate as the proximal ligand, and it appears that they also have the same overall mechanisms of dioxygen and C–H bond activation. The catalytic activity of P450 enzymes is thus determined by the generation of a short-lived, highly oxidizing intermediate at the active site heme iron, and the binding of a substrate molecule in close proximity for attack. Most bacterial P450s, and the eukaryotic enzymes involved in biosynthesis and degradation, have evolved to achieve high substrate specificity and regio- and stereo-selectivity of oxidation, which may be attributed to close complementarity between the substrate and the active site. In contrast, the detoxification enzymes such as the CYP1A and CYP2 subfamilies in the liver, and the CYP3A subfamily in the small intestine, have wide substrate ranges and catalyse less selective oxidation reactions [31]. It is likely that these enzymes have more open active sites, both in the sense of allowing easier substrate access and in having larger substrate pockets. As the activity of each P450 enzyme is controlled by the active-site architecture, it should in principle be possible to use site-directed or random mutagenesis to alter the characteristics of the active site pocket to change the substrate specificity and the selectivity of oxidation. An excellent example of this is the conversion of the mouse coumarin hydroxylase $P450_{coh}$ to an efficient testosterone 15α-hydroxylase by a single mutation [32, 33]. Although the amino acid sequences of $P450_{coh}$ and the mouse testosterone 15α-hydroxylase $P450_{15\alpha}$ differ by only 11 amino acids, the principle of modifying the active site to alter substrate specificity has been established.

1.3 The Scope of this Review

The purpose of this review is to examine the protein engineering of cytochrome $P450_{cam}$ with the aim of redesigning the active site to oxidize unnatural substrates, and to improve the activity and selectivity of the oxidation of organic molecules which are poor substrates for the wild-type protein. The focus is therefore on active site mutations, although other mutagenesis work will be discussed briefly in the context of the $P450_{cam}$ catalytic mechanism.

2 The Cytochrome $P450_{cam}$ System

The most well-characterised of all P450 enzymes is cytochrome $P450_{cam}$ from the soil bacterium *Pseudomonas putida* [34]. This readily isolated, stable and soluble enzyme catalyses the regio- and stereo-specific oxidation of camphor to 5-*exo*-hydroxycamphor. Almost every known physiochemical method of analysis has been applied to the study of this protein, which is also of great interest in theoretical and computational studies. The two electrons required for the monooxygenase reaction are derived from NADH, and passed to the $P450_{cam}$ hydroxylase via putidaredoxin reductase and putidaredoxin. All three of these proteins have been over-expressed in *E. coli* [35–37]. High-resolution crystal structures of $P450_{cam}$ are available, both in the substrate-free form [38] and with bound camphor [39], camphor analogues [40], and inhibitors [41, 42]. These factors combine to make the $P450_{cam}$ system an excellent model for genetic engineering-based investigations into many aspects such as the P450 catalytic mechanism, protein-substrate molecular recognition, protein-protein interactions and electron transfer, and the turnover of unnatural substrates for biotechnological applications. The mechanistic and structural aspects of $P450_{cam}$ activity are reviewed in this section.

2.1
The $P450_{cam}$ Catalytic Mechanism

2.1.1
Substrate Binding and Heme Reduction

The early steps and intermediates of the $P450_{cam}$ catalytic cycle are well established, and the most commonly proposed mechanism is shown in Fig. 1. In the ferric resting state and without bound substrate (**1**, Fig. 1), a hydrogen-bonded network of six water molecules fills the enzyme active site, one of which occupies the sixth heme coordination site [43]. The heme is in the low-spin state and its reduction potential is about –300 mV [44]. The reduction potential of putidaredoxin is about –240 mV, increasing to about –200 mV upon binding to $P450_{cam}$, and thus it clearly cannot reduce substrate-free ferric $P450_{cam}$.

The binding of camphor to $P450_{cam}$ results in the expulsion of all of the water molecules from the active site. The heme iron becomes five-coordinate (**2**, Fig. 1), its spin-state changes to predominantly high-spin, and the reduction potential increases to –170 mV so that putidaredoxin may now deliver the first electron and initiate the catalytic cycle. This interdependence of substrate binding, spin-state equilibrium and redox potential in $P450_{cam}$ has been discussed in detail elsewhere [45, 46]. The changes in spin-state and redox properties could be rationalised simply by the loss of the sixth ligand water; however recent quantum mechanical calculations by Harris and Loew [47] suggest that the presence of the water ligand alone is not sufficient to drive the heme of substrate-free $P450_{cam}$ to the low-spin state, and that the overall electrostatic field of the protein may also play an important role. The substrate-induced redox potential shift of $P450_{cam}$ is one of the most elegant examples of "gating" in biological electron-transfer, as reducing equivalents are not consumed unless the enzyme system is primed for substrate turnover.

The first electron-transfer from putidaredoxin to $P450_{cam}$ proceeds via a bimolecular complex [48] with a pseudo first-order rate constant of 10–15 s^{-1} [49]. The association constant was found to be highly ionic-strength dependent, suggesting an electrostatic interaction. Detailed investigations by Sligar, Poulos, Peterson and co-workers [50–52] led to the proposal that a group of basic surface residues (Arg-72, Arg-112, Lys-344, Arg-364), which lie directly above the heme proximal cysteine of $P450_{cam}$, interacts with acidic residues on the surface of putidaredoxin (possibly Asp-58, Glu-65, Glu-67, Glu-72, Glu77) to form the competent electron-transfer complex [53]. Mutations of Arg-72 and Lys-344 to glutamates resulted in slight decreases of protein-protein binding affinities [52], but mutations at Arg-112 substantially weakened $P450_{cam}$-putidaredoxin binding and reduced the electron transfer rates [54]. The C-terminus residue of putidaredoxin, Trp-106, has also been shown to be important in $P450_{cam}$-putidaredoxin recognition and electron-transfer [55].

The direct electrochemistry of $P450_{cam}$ at a bare edge-plane graphite (EPG) electrode has been reported recently [56]. The measured heme reduction potential of –286 mV was in close agreement with the previous spectroscopically determined value (-300 mV), and the camphor-induced shift in reduction

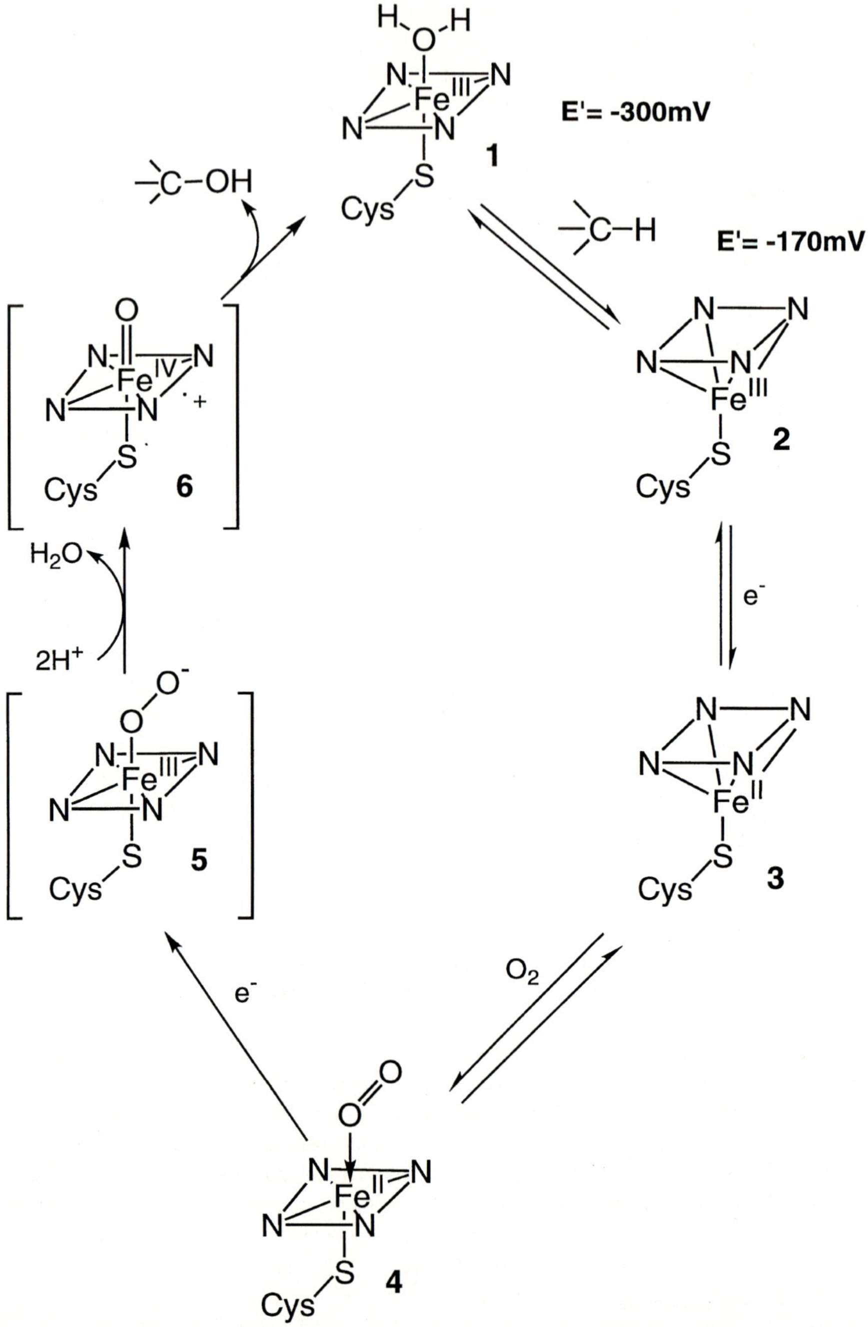

Fig. 1. The proposed catalytic cycle of $P450_{cam}$. The species in square brackets (**5** and **6**) have not been characterised but it is conventional to depict the active intermediate **6** as the iron(IV)-oxo porphyrin radical cation by analogy with horseradish peroxidase Compound I.

potential was also observed. It was proposed that the negatively charged EPG electrode surface mimicked the surface of putidaredoxin, and interacted with the basic residues on the $P450_{cam}$ heme proximal surface such that heterogeneous electron-transfer could occur.

2.1.2
Dioxygen and C–H Bond Activation

Oxygen binding to ferrous $P450_{cam}$ (**3**, Fig. 1) gives the ferrous-oxy form (**4**, Fig. 1), which is the last spectroscopically observable intermediate in the catalytic cycle. This species is unstable, with a half-life of approximately 90 s at 20°C [57, 58]. A second electron is then delivered by putidaredoxin. The rate constant of this step is $\approx 80\ s^{-1}$ at 4 °C, and thus at high putidaredoxin concentrations the first electron-transfer is the rate-limiting step in the catalytic cycle. At the lower physiological concentrations found in *P. putida*, however, the second electron-transfer step is rate-limiting.

The intermediates after the second electron-transfer have not been observed and the nature of the subsequent reactions remains controversial. The ferric-peroxide species (**5**) depicted in Fig. 1 is consistent with simple electron-counting. The mechanistic pathway thus far represents the heme-mediated two-electron reduction of a protein-bound dioxygen molecule to the peroxide. The proposed mechanism next calls for an O–O bond cleavage step leading to an oxy-ferryl species (**6**) and the formation of water. This is a two-electron step, reducing both oxygens in the peroxide. The two electrons required come from the heme, which is formally oxidised to $(Por)Fe^{V}=O$, and the two protons have to be "pumped" into the active site and delivered specifically to the distal peroxo oxygen and not to the heme-bound oxygen.

The ferryl species (**6**) is the commonly proposed active intermediate in P450-catalysed oxidation reactions, and is normally depicted as a $(Por)^{+}Fe^{IV}=O$ species by analogy with horseradish peroxidase Compound I, although its precise electronic structure has not been determined. This highly reactive intermediate abstracts a hydrogen atom from a substrate C–H bond, leading to a $(Por)Fe^{IV}$-OH species and a protein-bound substrate radical, which rapidly collapse to give the alcohol product and regenerate the ferric heme. This is known as the radical or oxygen rebound mechanism (Fig. 2) [59, 60]. There is a large body of evidence in support of the involvement of an intermediate in which the iron is formally in the +5 oxidation state [61, 62], and of a stepwise C-H bond oxidation mechanism [63]. However, other data, especially from heme model complexes [64–66] suggests that the O-O bond cleavage in a ferric-peroxide complex could proceed homolytically to give a $(Por)Fe^{IV}$–OH species and a hydroxyl radical which together oxidize a substrate C–H bond, whilst data from hydrogen peroxide-driven oxidation reactions catalysed by $P450_{LM2}$ have been interpreted on the basis of the ferric-peroxide species (**5**) being the active intermediate [67]. However, this evidence is not conclusive since none of these systems accurately reproduce the protein active site chemistry during normal catalytic turnover. On the other hand, convincing evidence from isotope labelling work indicates that the ferric-

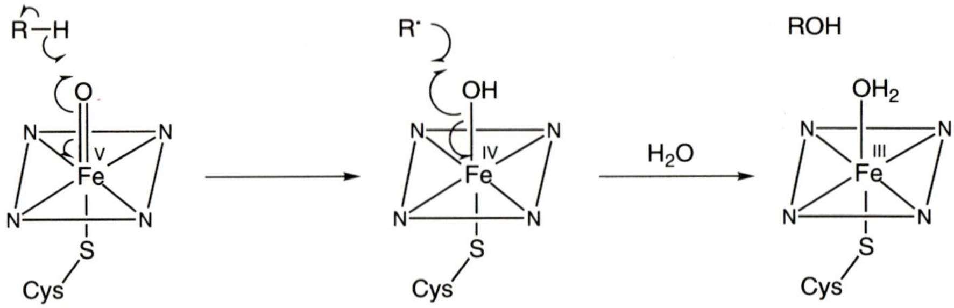

Fig. 2. The radical or oxygen rebound mechanism of C–H bond activation by P450 enzymes. The oxy-ferryl intermediate is shown formally as Fe(V) for electron counting purposes only, and the *half-arrows* indicate one-electron movements

peroxide acts as a nucleophile during aldehyde oxidation by some P450 enzymes [68, 69].

2.1.3 Uncoupling of the P450$_{cam}$ Catalytic Cycle

The oxidation of camphor by P450$_{cam}$ occurs with total stereospecificity, and all of the reducing equivalents from NADH are utilised for the formation of 5-*exo*-hydroxycamphor, i.e. the reaction has 100% coupling efficiency [70, 71]. The reactions with camphor analogues such as norcamphor and camphane exhibit reduced coupling (Sect. 3.2.3), and for molecules such as ethylbenzene and styrene which are structurally unrelated to camphor, the efficiency can be less than 5% [72, 73].

There are three possible pathways for channelling reducing equivalents away from the proposed P450 catalytic mechanism (Fig. 3) [74–76]. The first uncoupling pathway is the autoxidation of the ferrous-oxy form (**4**, Fig. 1), giving superoxide and regenerating the ferric resting state of the enzyme [57, 58]. However, the rate of this reaction is sufficiently slow that under normal turnover conditions it does not compete significantly with the other two uncoupling mechanisms or substrate oxidation [77]. The second pathway is the decomposition of the proposed ferric-peroxide intermediate (**5**, Fig. 1) to give hydrogen peroxide. The third uncoupling step is the two-electron reduction of the putative ferryl intermediate (**6**, Fig. 1) back to the ferric state, resulting in oxidase activity.

The second uncoupling pathway, that of peroxide generation, would compete significantly with substrate oxidation if the O–O bond cleavage step were inhibited, or if peroxide dissociation were promoted. Disrupting the proton delivery mechanism would inhibit O–O bond cleavage (see Sect. 3.2.3), while the rate of peroxide dissociation has been proposed to increase if the heme is accessible to water molecules during the catalytic cycle [77]. Water access to the heme could increase if the size and shape of the substrate were not sufficiently complementary to the P450$_{cam}$ active site, for example with some camphor analogues and those unnatural substrates which are smaller than camphor. The

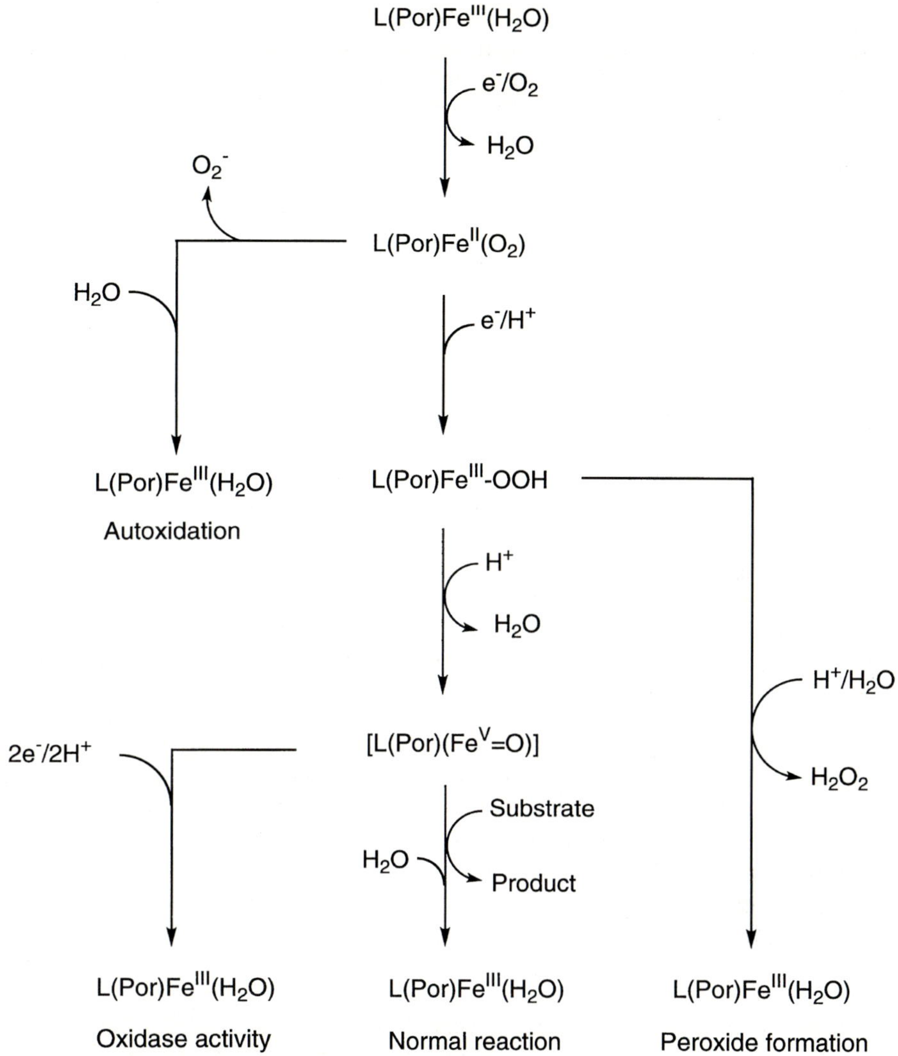

Fig. 3. The uncoupling mechanisms of P450 enzymes

exact mechanism by which water facilitates peroxide dissociation is not yet known, however, direct interaction with the heme iron or protonation of the heme-bound oxygen of the peroxide are the most likely possibilities.

The third uncoupling pathway is oxidase activity. The observation of this pathway is the most significant evidence in support of the mechanism outlined in Fig. 1, and the involvement of an intermediate in which the iron centre is formally Fe(V). Reduction of this species by putidaredoxin will compete with substrate oxidation if the substrate is positioned too far from the ferryl oxygen to undergo attack. This is particularly likely if the substrate is much smaller than camphor and is thus either mobile within the active site or it could be preferentially bound at a location far from the heme. Alternatively the substrate may be bound in an orientation such that no C–H bond is sufficiently close for attack, as in the case of 5,5-difluorocamphor [62].

With camphor as substrate, water is completely excluded from the active site during the catalytic cycle and the camphor is positioned with the C5 aligned for attack by the ferryl intermediate, leading to 100% coupling of oxidation.

2.2 Structural Aspects

2.2.1 *The Four Known P450 Crystal Structures*

High-resolution crystal structures have been determined for four P450 enzymes, $P450_{cam}$ [39], $P450_{terp}$ [78], $P450_{eryF}$ [79] and the heme domain of $P450_{BM\text{-}3}$ [80], all of which are bacterial in origin. Poulos [81], Deisenhofer [82] and Peterson [2] have all noted that the overall folds of the proteins are very similar despite the low sequence homologies (7–27%). A comparison of the structures showed that $P450_{terp}$, $P450_{eryF}$, and $P450_{BM\text{-}3}$ contained several large insertions compared with $P450_{cam}$ while, apart from the *N*-terminus, $P450_{cam}$ did not have any insertions relative to the other three. It was therefore proposed that $P450_{cam}$ represents the minimum P450 structure [82].

The amino acid side-chains which line the active sites of these four enzymes are shown in Fig. 4. Despite the similarities in the overall folds of the proteins, the active site structures, which largely determine the substrate specificity and hydroxylation selectivity, are substantially different. A substrate access channel is visible only in the structures of $P450_{BM\text{-}3}$ and $P450_{terp}$, and the other two proteins must therefore undergo either a reversible conformational change or dynamic structural fluctuations to allow the substrate to enter the active site. $P450_{cam}$ and $P450_{eryF}$ have clearly defined hydrophobic cavities of distinctly different size and shape, each complementary to its natural substrate and possessing highly specific protein-substrate interactions. Of particular note are the hydrogen bonds between Y96 and the camphor carbonyl in $P450_{cam}$, and the water-mediated hydrogen bonds between Y75, N89, and the backbone carbonyl of A241 and the 6-deoxyerythronoloide substrate in $P450_{eryF}$. On the other hand, $P450_{BM\text{-}3}$ and $P450_{terp}$ have channels extending from the surface to the heme, with the particularly long channel in $P450_{BM\text{-}3}$ presumably reflecting its specificity for long-chain fatty acids. It may be noted that even the most sophisticated sequence alignment and structure modelling methods could not have predicted the precise active site structure of one of these proteins from the other three structures.

2.2.2 *The Structure of $P450_{cam}$*

The structure of $P450_{cam}$ consists of two distinct domains, one helix-rich and the other helix-poor [39]. The residues which line the active site come from both domains, and there are extensive contacts with the bound camphor (Figure 5). Poulos observed that the active site is highly hydrophobic, and that the crucial residues involved in camphor binding are F87, Y96, T185, L244, V247, and V295. The *gem*-dimethyls of the V295 side-chain interdigitate with the C8 and C9 *gem*-dimethyls of camphor (Fig. 6), the V247 side-chain contacts C6 of the camphor molecule, V244 contacts the C1 methylene and C10 methyl groups, while T185 contacts the C10 methyl group. The side-chain of F87 is in van der Waals contact with the camphor carbonyl oxygen, which forms a hydrogen bond with the phenolic side-chain of Y96. These non-covalent contacts are individually weak, but together they contribute much to the energetics of camphor binding. It was proposed that the hydrogen bond between Y96 and the camphor carbonyl was mainly responsible for orientating the substrate for specific hydroxylation and contributed little to substrate binding (See Sect. 3.2.3). Thus in the absence of substrate, the hydroxy group of Y96 interacted with water molecules present in the active site, and the enthalpy gained from the formation of the hydrogen bond with the camphor carbonyl would be balanced by the enthalpy required to expel these waters. It follows therefore that the entropy gain arising from transferring camphor from the aqueous medium to the hydrophobic protein substrate pocket, and the release of the active site waters, is a major driving force for camphor binding [87].

An unusual deformation in the active site region of the I-helix gives rise to a pocket above the heme which has been proposed as a possible binding site for oxygen [39]. In the structure of carbonmonoxy-P450$_{cam}$, Raag and Poulos [88] found that the bound CO projected into this pocket. Assuming that a dioxygen molecule would also be bound in a similar orientation, it was proposed that the hydroxy group of T252 could stabilise dioxygen binding in P450$_{cam}$ by forming a hydrogen bond to the distal oxygen atom. Furthermore, a number of water molecules were found in the immediate vicinity of this proposed oxygen binding pocket in the camphor-bound structure. These waters were hydrogen-bonded to various amino acid residues, and thus the side-chain of T252 could also be involved in the transfer of protons required for O–O bond cleavage [39].

3 Protein Engineering of P450$_{cam}$

3.1 General Considerations

The four known crystal structures of bacterial P450s clearly show the diversity in active site architecture of this superfamily of enzymes. Each of these enzymes has evolved to achieve a high degree of specificity towards a particular target molecule and high regio- and stereo-selectivity of oxidation, and they could therefore be expected to exhibit markedly different active-site topologies. This suggests that, in general, the substrate binding pockets of P450 sub-families which target different classes of molecules will be very different. As the first P450 enzyme for which the X-ray crystal structure was determined, P450$_{cam}$ has been widely used as a prototype in molecular modelling studies, and three-

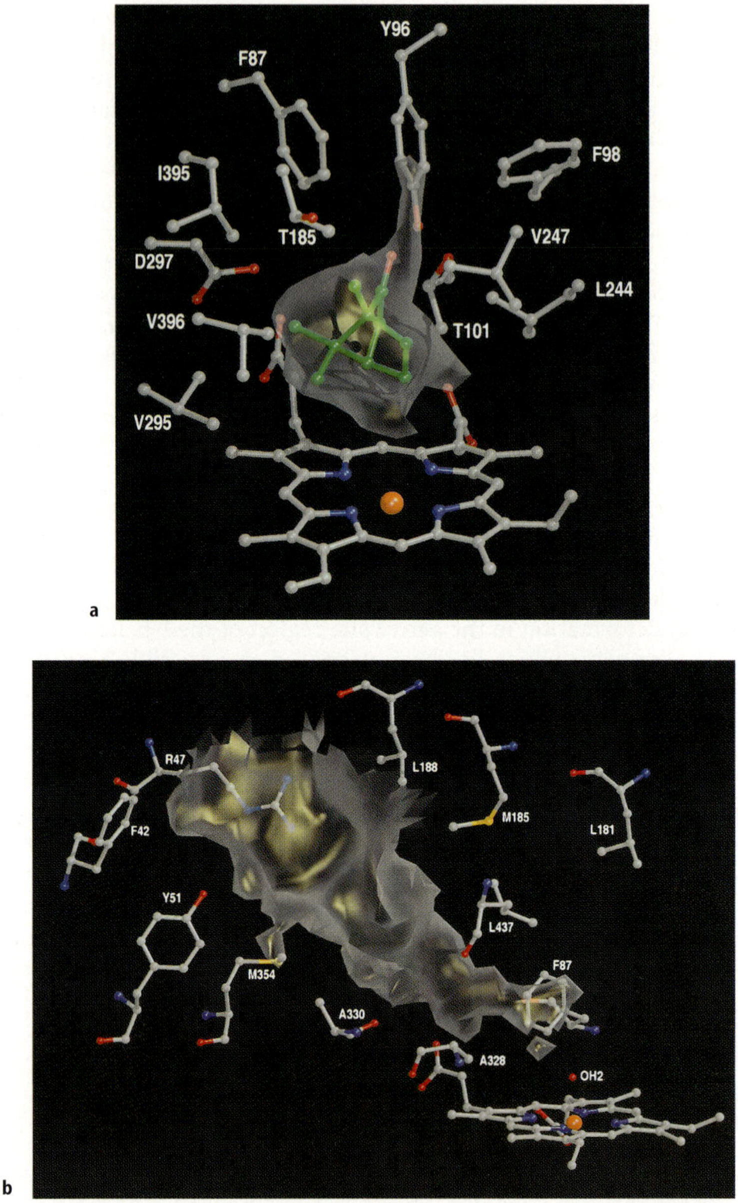

Fig. 4. The active site structures of **a** P450$_{cam}$, **b** the heme domain of P450$_{BM3}$, **c** P450$_{terp}$, and **d** P450$_{eryF}$, including bound substrates when observed, and showing the solvent-accessible area of each protein. GRASP [83] was used to calculate maps of the solvent-accessible surfaces of each protein from deposited Brookhaven Protein Data Bank (PDB) coordinates, using a probe radius of 1.4 Å and excluding the substrate molecule, if present. Cavities and channels were interactively assigned from within GRASP, combined with Molscript [84] representations of the active sites (including substrate) in appropriate orientations, and collectively rendered by Raster3D [85, 86]

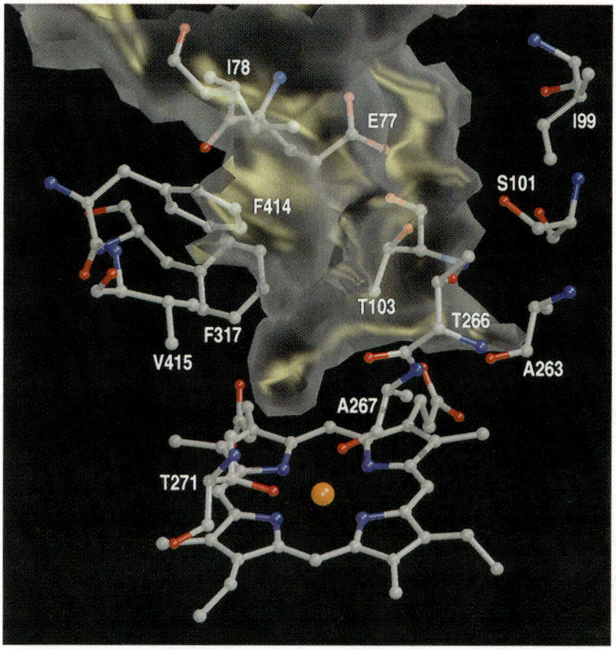
I78
E77
I99
S101
F414
T103
T266
F317
V415
A263
A267
T271

c

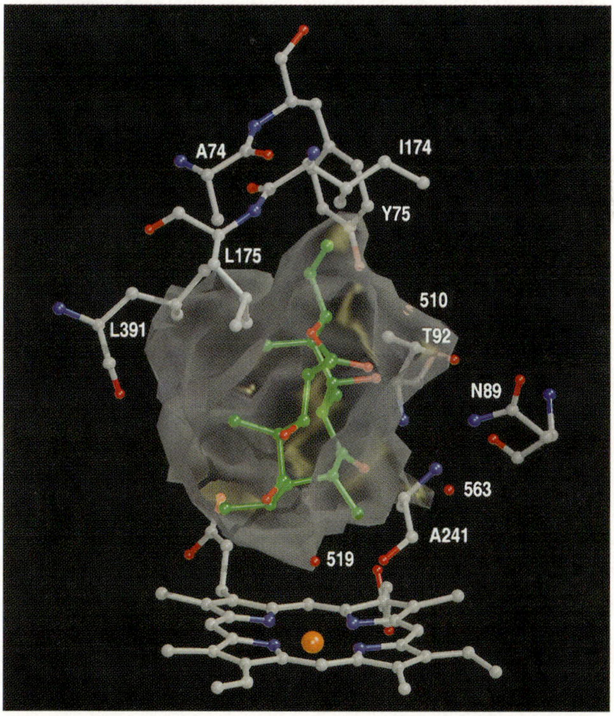
A74
I174
Y75
L175
510
L391
T92
N89
563
A241
519

d

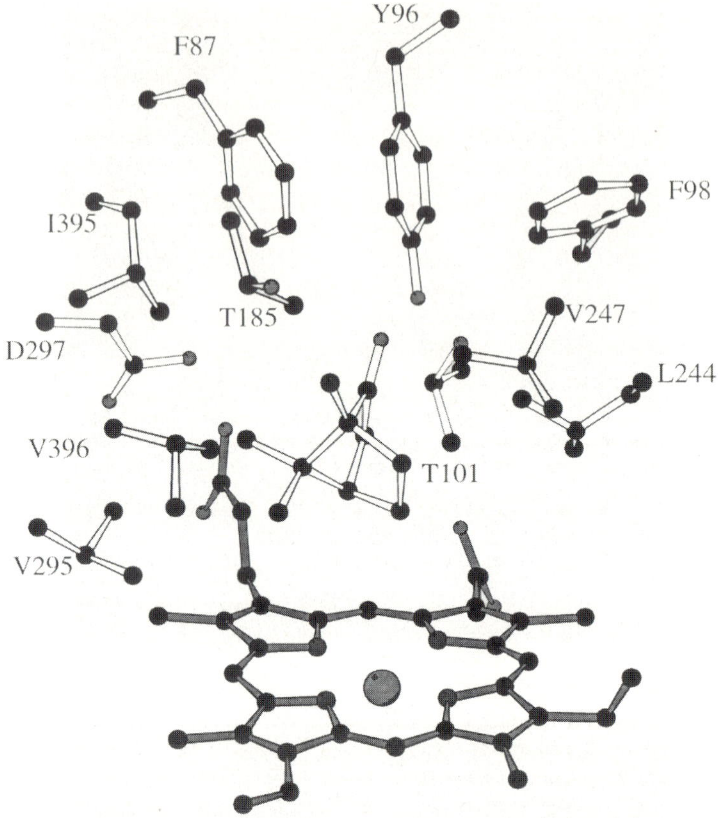

Fig. 5. The active site structure of camphor-bound $P450_{cam}$ showing the residues which contact the camphor substrate

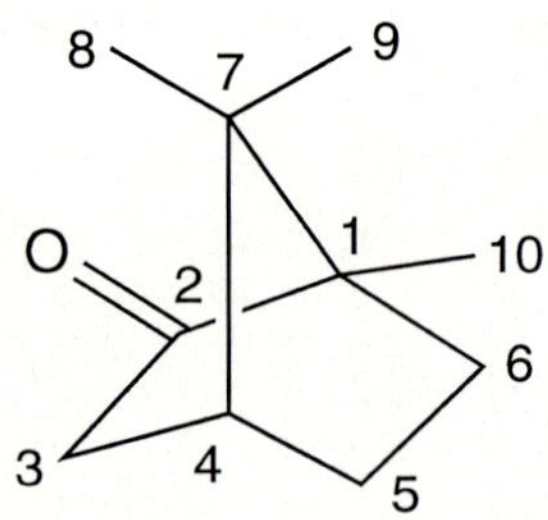

Fig. 6. The numbering of carbon atoms in camphor

dimensional structural models based upon P450$_{cam}$ have been used to predict possible protein-substrate binding interactions for a variety of P450s [89–92]. Since the publication of the X-ray structures of P450$_{terp}$ and P450$_{BM-3}$, more sophisticated approaches have been developed which generate models based upon all of the available structures [93–96]. However, an important finding has been that the exact details of the structure obtained, and particularly the identity and position of active-site residues, is dependent upon the methods used in generating the model [97, 98]. Indeed it must be stressed that whilst such studies are capable of predicting gross architectural features of P450 enzymes and their likely active site residues, the present structural database is too small, and the sequence homologies too low for precise protein-substrate interactions to be elucidated using these techniques [99].

Biotechnological applications such as drug metabolism and toxicity screening require the use of human P450 enzymes. However, in other areas such as biotransformation and environmental bioremediation, genetically engineered forms of the stable, soluble bacterial enzymes will have important applications. Many precursors to useful synthetic intermediates and pharmaceutical compounds are not substrates of naturally occurring P450 enzymes, and even in favorable instances where they are, the substrate hydroxylation reaction may not have the required regio- or stereo-selectivity. High selectivity will be obtained if the substrate can be orientated within the active site of an engineered enzyme such that the target C–H bond is closest to the ferryl oxygen. It is also desirable to have high coupling efficiency of substrate oxidation. An important step towards the use of engineered P450 enzymes in these applications is therefore a better understanding of the factors which control the substrate range and the catalytic activity, selectivity and coupling of oxidation, all of which could be altered by changing the topology of the active site by mutagenesis. A useful approach to this problem is a combination of protein and substrate engineering. The increasingly sophisticated and reliable methodologies of computer simulation and substrate docking calculations will also make important contributions. For example, Montellano has recently reported the computer screening of 20000 organic molecules for their likelihood of oxidation by wild-type P450$_{cam}$ [100]. Of the eleven compounds which were predicted to be substrates for wild-type P450$_{cam}$, seven induced spin-state shifts and eight showed increased NADH consumption rates and were therefore strong candidates as substrates.

3.2
Active Site Mutations

In this section, active site mutations designed to investigate the mechanistic details of P450$_{cam}$ catalysis are discussed. In a series of seminal papers, a combination of genetic engineering and detailed investigations into substrate range and turnover activity, in some cases combined with X-ray crystal structure determination, has provided insights into this fascinating area, and the results lay down an excellent basis for rational protein redesign [27, 101].

3.2.1
Substrate Access

The crystal structures of $P450_{cam}$ with and without bound camphor revealed no obvious channel for substrate molecules to diffuse into the active site, suggesting that protein conformational changes would be necessary for this to occur. A detailed comparison of the two structures by Poulos [37, 43] identified regions of the protein with higher thermal mobility in the substrate-free form, including the residues 87–96 in the B′ helix, residues 185–193, and D251 which forms a salt-bridge with K178. Molecular dynamics calculations have also suggested that the F87 side-chain is highly mobile [102]. The crystal structure of $P450_{cam}$ with a bound inhibitor which contained a C_8 alkyl group showed that this aliphatic chain pointed away from the heme and out towards the protein surface [42], and that the side-chains of F87, Y96 and F193 moved far away from their former positions in the substrate-free structure. In the case of F193, the side-chain moved onto the protein surface, and it was proposed that this residue might serve as an initial substrate recognition site and then chaperone the target molecule into the active site.

The roles of F87, Y96 and F193 in substrate access have been investigated by site-directed mutagenesis [27]. The Y96F mutation had little effect on either k_{on} or k_{off} of camphor binding, the F87W mutation reduced k_{on} by a factor of 6, while the F193C mutation did not significantly affect k_{on} but increased k_{off} 5-fold. This data strongly supports the proposal that the F87 and F193 residues each form part of the substrate access channel.

3.2.2
Active Site Topology Investigations

Montellano and co-workers have developed a chemical method of mapping the active site topology of P450 enzymes [103, 104]. The product distribution of aryl migration from iron to pyrrole-nitrogens in an aryl-heme complex gives an indication of the space available above each of the four pyrrole rings in the active site. This method has been applied to wild-type $P450_{cam}$ and the F87A and F87W active site mutants. In the crystal structure of wild-type $P450_{cam}$, the F87 side-chain lies over one of the pyrrole rings and also contacts the bound camphor (Fig. 5). The oxidation of camphor by both mutants gave 5-*exo*-hydroxycamphor as the only product, suggesting that the crucial protein-substrate contacts which impart the regio- and stereo-specificity of camphor oxidation were maintained.

In the aryl migration experiments, it was found that whilst increasing the size of the side-chain in the F87W mutant had little effect, the pyrrole nitrogen covered by F87 was much less accessible in the F87A mutant. Since the F87A mutation appeared to result in less, rather than more room being available, it was concluded that there was a local structural rearrangement, or falling-in, of the active site when the small methyl side-chain of alanine was introduced. Such a collapse of the active site upon mutagenesis is potentially a problem in $P450_{cam}$ genetic engineering, and in protein engineering work in general.

However, as the F87A mutant retained the stereospecificity of camphor oxidation, it was suggested that the F87 side-chain was high enough in the active site pocket for the local structural changes not to affect the binding orientation of camphor close to the heme.

3.2.3
Oxidation of Camphor and Analogues

The role of active site residues in camphor binding and oxidation has been investigated by Sligar and co-workers. In the crystal structure of camphor bound P450$_{cam}$, the side-chain of Y96 formed a hydrogen bond with the camphor carbonyl [39]. The Y96F mutation was constructed to remove this hydrogen bond whilst conserving the rest of the active site structure. Camphor binding to the Y96F mutant induced a heme spin-state shift to 59% high-spin (HS) compared with >95% HS with the wild-type [105]. This change in spin-state shift could arise from greater mobility of camphor in the mutant, thereby allowing solvent water molecules to enter the active site and bind to the heme iron [106]. The camphor dissociation constant was only slightly raised in the mutant. Hence, as predicted by Poulos [39], the hydrogen bond between the side-chain of Y96 and camphor contributed little to the overall thermodynamics of substrate binding. The selectivity of camphor hydroxylation was slightly reduced, with 5-*exo*-hydroxycamphor constituting 92% of the products, and the reaction remained tightly coupled (Table 1). The largest difference was observed in the turnover activity – the NADH turnover rate of the mutant was only 50% that of the wild-type, and the k_{cat}/K_m efficiency ratio was 5-fold lower.

The effect of protein/substrate hydrogen bonding was also investigated using camphor analogues (Table 1) [105]. Thiocamphor oxidation was tightly coupled to NADH consumption, but the selectivity for the 5-*exo* position was reduced to

Table 1. **a** Effect of specific protein-substrate contacts on the activity of wild-type P450$_{cam}$. **b** Effect of mutations of F87 and Y96 on the oxidation of camphor and analogues

	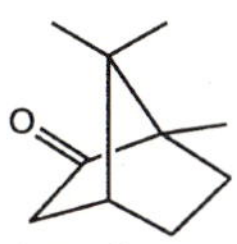	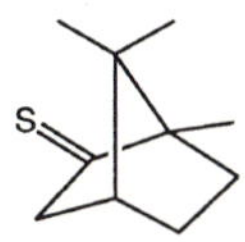	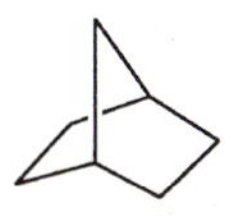
a Wild-type P450$_{cam}$	Camphor	Thiocamphor	Camphane
% HS	> 95	65	46
% coupling	100	98	8
% 5-*exo*-OH	100	64	90
b Camphor oxidation	Y97F	F87A	F87W
% HS	59	60	100
% coupling	100	–	–
% 5-*exo*-OH	92	100	100

65 %. X-ray crystallography however showed that the substrate was bound in an unusual orientation, with the sulphur pointing towards the iron [40]. Camphane, which lacks the carbonyl group as well as the methyl substituents of camphor, was hydroxylated with 90 % selectivity at the 5-*exo* position, but the reaction was highly uncoupled. These results demonstrate that the Y96-camphor carbonyl hydrogen bond is important in achieving high catalytic activity and coupling efficiency, but it appears that other factors may also be important in substrate binding and in the control of the regioselectivity of hydroxylation.

Sligar and co-workers have examined the role of non-covalent contacts between camphor and the $P450_{cam}$ active site residues in the monooxygenase reaction. The side-chains of the V247 and V295 residues are close to the heme and in van der Waals contact with camphor methyl groups. V247 contacts the camphor C6 methylene and C10 methyl groups, while V295 contacts the C8,C9 *gem*-dimethyls (Fig. 5). These interactions were investigated by studying the selectivities of the wild-type and the V247A and V295I mutants towards the oxidation of camphor and the analogues 1-methylnorcamphor and norcamphor [107]. The results are summarized in Table 2. The data for camphor oxidation with these two mutants, and with the Y96F, F87A and F87 W mutants shows the remarkable insensitivity of the stereoselectivity and coupling of camphor oxidation to single site mutations, even at residues in direct contact with the substrate.

Altering the substrate by removing the camphor methyl groups significantly changed the oxidation selectivity and coupling. It is also clear from the data that the spin-state shift is not a reliable indicator of either the selectivity or the coupling efficiency of substrate oxidation. The stereoselectivity of 1-methylnorcamphor and norcamphor oxidation by wild-type was substantially lower

Table 2. The effect of variations in non-covalent contacts on the monooxygenase activity of $P450_{cam}$.

		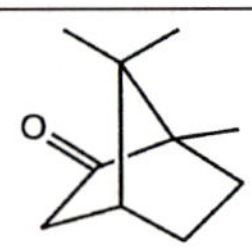 Camphor	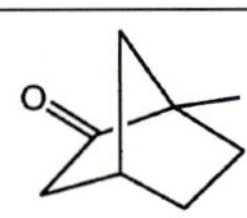 1-Methylnorcamphor	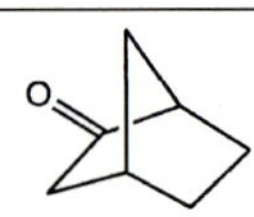 Norcamphor
WT	% HS	> 95	48	45
	% coupling	100	45	12
	% 5-*exo*-OH	100	82	45
V247A	% HS	95	37	44
	% coupling	100	40	10
	% 5-*exo*-OH	97	72	40
V295I	% HS	–	49	–
	% coupling	–	45	–
	% 5-*exo*-OH	100	90	46
Y96F	% HS	59	–	–
	% coupling	100	–	–
	% 5-*exo*-OH	92	63	36

than for camphor, and was reduced even further in the V247A mutant. However, the selectivity of the V295I mutant was higher than that of the wild-type, presumably due to partial restoration of the protein-substrate van der Waals contact by lengthening the 295 side-chain in the mutant such that it could interact with the C7 methylene of 1-methylnorcamphor. It is also interesting to note that although the Y96F mutation did not greatly affect the selectivity of camphor oxidation, its importance became evident when other complementary non-covalent contacts were removed. This work provides excellent examples of the cumulative effect of individually weak protein-substrate interactions, and of how mutagenesis can improve the selectivity of the wild-type protein towards a particular substrate molecule.

3.2.4
O–O Bond Cleavage

The side-chain of T252 is proposed to stabilise the ferrous-oxy form and mediate the delivery of the protons required for dioxygen activation from the external medium into the active site [39]. Martinis et al. [108] and Imai et al. [109] have investigated the effect of mutations at this position. All the mutants gave 5-*exo*-hydroxycamphor as the only product, but there were large differences in the coupling efficiencies. When T252 was substituted by amino acids with hydrophobic side-chains, camphor hydroxylation was highly uncoupled, and the coupling efficiency increased with side-chain size: 3% for T252G, 6% for T252A, and 22% for T252V. However, mutants with polar or charged residues such as Ser and Lys showed high couplings of up to 80%. Furthermore, Kimata et al. reported that the mutant in which threonine was substituted with an unnatural amino acid where the OH group was replaced with an OMe group showed 100% coupling efficiency of camphor hydroxylation, although the turnover activity was reduced [110]. It was concluded that a free hydroxy group at the 252 position was not absolutely required for the monooxygenase reactivity of P450$_{cam}$, although a polar or charged side-chain was clearly beneficial.

The catalytic properties of the T252A mutant have been studied in detail [108]. The reducing equivalents were channelled to give 6% product, 51% hydrogen peroxide, and 44% water. The crystal structure of this mutant determined by Poulos and co-workers [111] showed that local structural rearrangements in the vicinity of the mutation resulted in the opening of a water access channel from the solvent into the active site. A new water molecule was also located in the active site and found to be in contact with camphor. This increased water access to the active site could explain the increased uncoupling via peroxide formation.

Based on kinetic solvent isotope effect data, Aikens and Sligar proposed a charge-relay mechanism in which the side-chains of D251 and T252 were involved in hydrogen bonding, either directly or via a water molecule, with the distal oxygen of the ferric-peroxide intermediate, leading to proton delivery in the O-O bond cleavage step [112]. Taking into account the results with the unnatural amino acid substitution at T252, the latter mechanism involving an extra water molecule appears to be more likely. Gerber and Sligar have examined the effect of the D251N mutation on camphor oxidation [113, 114]. This mutant showed a 100-fold decrease in NADH turnover activity compared to the wild-type, although camphor oxidation remained tightly coupled to NADH consumption. A detailed analysis of the individual steps leading to camphor hydroxylation suggested that the rate-limiting step had changed to O–O bond scission rather than the electron-transfer steps. These observations are consistent with the intimate involvement of the D251 side-chain in the O–O bond cleavage step, most likely in proton delivery from the external solvent specifically to the distal oxygen of the ferric-peroxide intermediate. Shimada et al. showed that mutations at K178 and R186, which are near the protein surface but also in close proximity to D251, also affected the catalytic activity [115]. A pro-

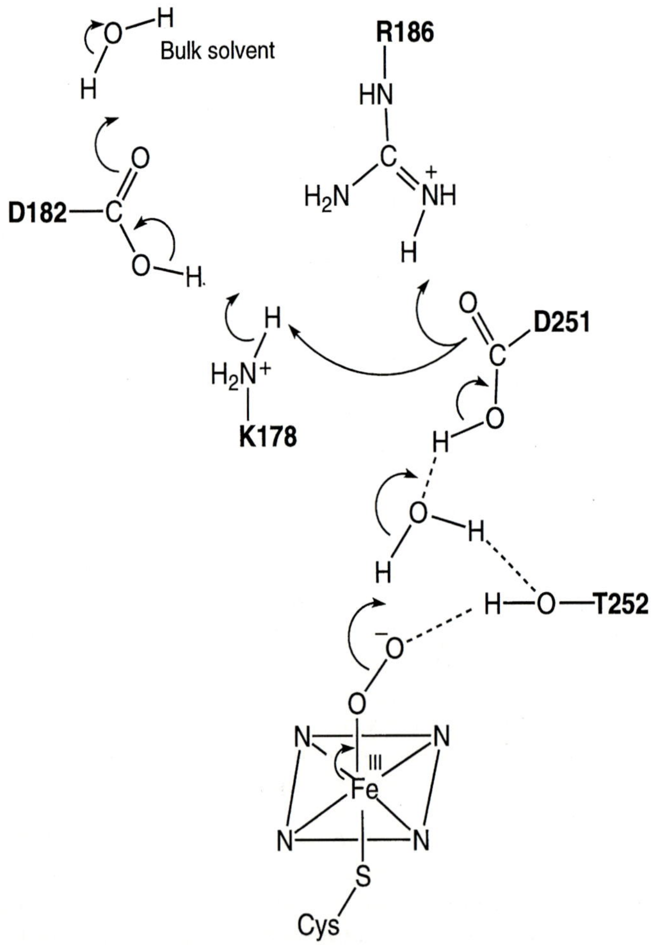

Fig. 7. The proposed proton delivery pathway from the solvent medium specifically to the non-heme-bound oxygen of the ferric-peroxide intermediate and the mechanism of O–O bond cleavage, with the arrows indicating the movement of electron pairs

posed mechanism, which takes into account all these data, of proton delivery and O–O bond activation is shown in Fig. 7 [110, 112].

3.3 The Oxidation of Ethylbenzene and Alkanes

The active site of $P450_{cam}$ has many complementary contacts with the camphor molecule, and the substrate range of this enzyme is quite narrow when compared to P450 enzymes involved in detoxification. Apart from the camphor analogues discussed above and some related molecules such as adamantane and adamantanone, wild-type $P450_{cam}$ also has low activity towards the oxidation of ethylbenzene [72, 77, 116], styrenes [73, 117], a tetralone derivative [118], nicotine [119], and tetrahydronaphthalene [120]. The oxidative and reductive dehalogenation of haloalkanes have also been described [121–123].

3.3.1 *Protein Engineering Studies on Ethylbenzene Oxidation*

Loida and Sligar have reported the most detailed investigation into the use of site-directed mutagenesis to increase the activity and selectivity of $P450_{cam}$ for the oxidation of an unnatural substrate [77]. Mutations at active site residues located at different distances from the heme were investigated for their effect on the turnover activity and the partitioning between the uncoupling pathways in the oxidation of ethylbenzene. Wild-type $P450_{cam}$ regiospecifically hydroxylated ethylbenzene at the benzylic carbon [72], and the reaction was highly uncoupled, with the majority of the NADH consumed (80%) being channelled to peroxide generation and only 5% being utilized for product formation [116]. The 1-phenylethanol product was formed with some stereoselectivity, with an *R*:*S* ratio of 73:27 (46% enantiomeric excess, ee).

The rational active-site redesign was based on the important observation that the active-site residues are ordered in nearly parallel planes or tiers relative to the heme group [77]. The residues are: Tier 1 – T101, L244, G248, V295, D297; Tier 2 – F87, Y96, V247, I395, V396; Tier 3 – T185. It was proposed that introducing residues with larger side-chains into Tier 1 would probably block substrate access to the ferryl intermediate and hence reduce the coupling efficiency of substrate oxidation, while introducing bulk in Tiers 2 and 3 should force the substrate closer to the heme and increase the coupling. These proposals are strongly supported by the results of the mutagenesis experiments summarized in Table 3. It was noted that most mutations which introduced bulkier side-chains and thus increased the hydrophobicity of the active site reduced uncoupling via peroxide formation, consistent with water access being a key factor in this pathway.

The partitioning between uncoupling via oxidase activity and substrate oxidation at the ferryl intermediate is a potential probe of the substrate binding orientation. Active site mutations resulted in large changes in the partitioning between the two pathways, such that in the T101M mutant the pathways were biased 99.6% in favor of oxidase activity while in the T185F mutant 65% of the

Table 3. Effect of $P450_{cam}$ active site mutations on the catalytic profile of ethylbenzene oxidation

$P450_{cam}$ protein	% HS	NADH turnover rate[a]	1-phenyl-ethanol formation rate[a]	% coupling[b]	% H_2O_2[b]	% H_2O[b]
Wild-type	33	52	2.62	5	77	11
T185V	–	49	2.45	5	76	13
T185L	24	64	8.64	13	72	12
T185F	4	11	1.43	13	76	7
V247A	35	30	0.90	3	4	48
V247M	31	40	4.80	12	67	13
T101M	5	14	0.02	0.1	61	24
T101I	–	26	0.57	2.2	83	10
V295I	25	41	0.12	0.3	68	17
T101M-T185F-V247M	5	23	2.99	13	52	20

[a] Rates given as nmol (nmol $P450_{cam}$)$^{-1}$ (minute)$^{-1}$.

[b] Given as the percentage of reducing equivalents from NADH channelled to the formation of product (% coupling), hydrogen peroxide or water.

ferryl intermediates attacked ethylbenzene. The results are entirely consistent with substrate positioning being the key factor in determining the coupling efficiency at the ferryl intermediate. Increasing the side-chain volumes in the Tier 1 mutants T101M and V295I all but eliminated product formation, due to steric hindrance with the bulkier side-chains blocking substrate access to the ferryl. Increasing the bulk in Tier 2 by the V247M mutation increased product generation relative to oxidase activity, while the V247A mutation had the opposite effect. A Tier 3 substitution such as T185V, which does not significantly alter the side-chain volume, had little effect on the catalytic profile. However, increasing the side-chain volume in the mutants T185L and T185F progressively increased the partitioning towards product formation. These two mutations decreased the space in the upper region of the active site and forced the substrate towards the heme, thus increasing the efficiency of substrate oxidation.

Based on all the data, the triple mutant T101M-T185F-V247M was constructed and it was shown that the effects of the mutations were additive, with the detrimental effects of the T101M mutation being more or less reversed by reducing the active-site volume higher up in the pocket. This additivity also implies that these three mutations at different locations in the active site induced minimal structural rearrangements.

Although the general trends in the data may be explained in terms of changing the bulk in each tier, a number of points may also be noted. Firstly, the T185L mutation increased the coupling primarily by retarding the peroxide pathway, whilst the T185F mutation principally lowered the oxidase activity. This data cannot be rationalised on steric grounds alone, and suggests that the substrate-binding orientation is also important. Furthermore, whereas the Tier 2 mutation V247M reduced uncoupling via the peroxide pathway, the V247A mutant showed a dramatic switch almost completely to oxidase activity. It was proposed either that structural rearrangements may have actually decreased the active site volume in the V247A mutant, or that the proton transfer pathway may have been perturbed. However, a decreased volume in Tier 2 would be expected to lead to the binding of ethylbenzene closer to the ferryl intermediate and thus reduce rather than increase oxidase activity, implying that the proposed rearrangements would somehow have to result in decreased volume within Tier 1. These details notwithstanding, this seminal work demonstrates that water and substrate access to the heme, and hence the coupling efficiency in the oxidation of unnatural substrates, may be rationally predicted and altered.

3.3.2
Rational Protein Redesign for the Oxidation of Alkanes

Active site mutants of $P450_{cam}$ have been shown to have high activity for the oxidation of small ($C_5 - C_7$) linear and branched alkanes. As these target molecules are highly hydrophobic, it was postulated that replacement of the most hydrophilic active-site residue, Y96, with phenylalanine or alanine should promote their binding and catalytic oxidation. The catalytic parameters of alkane oxidation by the Y96A and Y96F mutants are compared with those of the wild-type $P450_{cam}$ in Table 4 [124]. It was found that with linear alkanes as substrates, the NADH consumption rates of the mutants were up to 19 times that of the wild-type. Furthermore, substantially increased coupling of substrate oxidation to NADH consumption meant that the product formation rates were up to 275 times that of the wild-type. With branched alkanes, the differences were smaller, although again the mutants were more active than the wild-type and showed higher coupling. These results confirmed that the increased active-site hydrophobicity of the mutant enzymes substantially promoted alkane oxidation.

Of particular interest was the observation that for each alkane studied, the coupling efficiencies of the three enzymes were in the order Y96F > Y96A ≫ wild-type. This was rationalised in terms of the increased hydrophobicity of the substrate binding pocket in the mutants, which should exclude water more effectively from the active site during catalysis, and so inhibit protonation of the heme-bound oxygen in the ferric-peroxide intermediate with subsequent release of hydrogen peroxide.

The product distributions observed were in accordance with the radical rebound mechanism of C–H bond activation. Thus the most reactive tertiary bonds were attacked in preference to secondary bonds, with no hydroxylation at primary positions, suggesting that the alkanes are somewhat mobile in the active site, with the proteins exerting little control over their binding orientations. There was also little variation in product distribution between the wild-type and mutant proteins. The Y96 residue is located at the top of the substrate pocket and far-removed from the heme, and thus varying the side-chain volume from the Y96A to the Y96F mutant may not significantly affect the binding orientation of alkanes in the vicinity of the heme. The lack of regioselectivity of the engineered $P450_{cam}$ mutants is unlike natural alkane hydroxylases, which

Table 4. The NADH turnover and total product formation rates and derived coupling efficiency of alkane oxidation catalyzed by wild-type, Y96A and Y96F mutant $P450_{cam}$.

Alkane substrate	NADH turnover rate[a]			Total product formation rate[b] (% Coupling)[c]		
	WT	Y96A	Y96F	WT	Y96A	Y96F
Pentane	34.8	419.6	404.8	0.9 (2.6)	60.8 (14.5)	105.7 (26.1)
Hexane	19.5	388.7	234.4	0.4 (2.0)	89.3 (23.0)	109.8 (47.0)
Heptane	7.9	103.4	55.9	0.2 (2.5)	25.7 (25.0)	18.3 (32.7)
2-methylpentane	77.1	210.9	222.0	5.9 (7.7)	86.0 (40.8)	97.0 (43.7)
3-methylpentane	130.4	191.2	224.0	27.7 (21.3)	99.1 (52.0)	122.6 (54.7)
2-methylhexane	31.6	90.7	57.8	1.7 (5.3)	18.9 (20.8)	17.9 (30.9)

[a] Given as nmole NADH consumed (nmole $P450_{cam}$)$^{-1}$ (minute)$^{-1}$.
[b] The total amount of alkane oxidation products formed (nmole) (nmole $P450_{cam}$)$^{-1}$ (minute)$^{-1}$.
[c] The coupling is the ratio of the total amount of products formed to the amount of NADH consumed and expressed as a percentage.

generally catalyse ω-C–H bond hydroxylation. This difference in activity could be attributed to different active site topologies, with the natural alkane hydroxylases perhaps having more channel-like substrate-binding pockets and less space in the vicinity of the hydroxylation centres, such that only terminal methyl groups can approach close enough to the ferryl for attack.

3.4 Investigations into the Selectivity of the Oxidation of Phenyl Derivatives

The substrate recognition interactions and selectivity of C-H bond activation by $P450_{cam}$ have been further studied by investigating the oxidation of a number of phenyl derivatives by the wild-type and the active site mutants Y96G, Y96A, Y96V, Y96L and Y96F [125, 126]. These mutations were designed to create active sites of varying sizes which were more hydrophobic than that of the wild-type enzyme. The compounds diphenylmethane and benzylcyclohexane probed the selectivity for aliphatic vs. aromatic C–H bond activation and the effect of a flexible cyclohexane ring against the rigidity of a phenyl group. The selectivity for C–H bond oxidation in the presence of reactive functionalities was investigated with the compounds diphenylamine, 1,1-diphenylethylene and 1-phenyl-1-cyclohexylethylene. An important aspect of this study was that the reconstituted $P450_{cam}$ system was found to be sufficiently stable and active to generate enough products for purification by HPLC, and identification by chemical methods such as NMR and mass spectroscopy.

Wild-type $P450_{cam}$ was not active towards any of these compounds (Table 5), all of which are significantly larger than camphor. The diphenyl compounds were oxidized only by the Y96G and Y96A mutants, while the mixed ring compounds were also attacked by the other mutants which should have smaller active sites. It was proposed that the cyclohexane ring can adopt many different conformations distinct from the most stable chair form, thus enabling the mixed ring substrates to minimise steric hindrance with the active site side-chains.

The oxidation of diphenylmethane, diphenylamine and 1,1-diphenylethylene by the Y96A and Y96G mutants was regiospecific, giving *para*-hydroxylated products, suggesting that these substrates were bound in very similar orientations. It was proposed that the Y96A and Y96G mutations generated extra space in the active site to accommodate one of the phenyl rings of these substrates. This proposed binding orientation would position the *para*-C–H bond of the second phenyl group closest to the ferryl intermediate, thus giving rise to regiospecific hydroxylation without attacking the amine group or olefinic double bond.

Benzylcyclohexane was oxidized mainly to *trans*-4-benzylcyclohexanol (Fig. 8), and the selectivity decreased with increasing side-chain volume at the 96 position. The oxidation of 1-phenyl-1-cyclohexylethylene also occurred exclusively on the aliphatic ring, with attack at C4 being predominant, and no attack at more activated C-H bonds or the olefinic functionality (Figure 8). The presence of aliphatic side-chains such as those of L244, V247 and V295 in the immediate vicinity of the heme may give rise to stronger van der Waals interactions with the cyclohexane ring than with the planar phenyl group, thus preferentially positioning the cyclohexane ring near the heme. The product distributions also showed some interesting trends. The Y96A and Y96G mutants show high diastereoselectivity in the oxidation of benzylcyclohexane, giving 90% of the *trans* diastereoisomer (Fig. 8). The reduced diastereoselectivity observed with the Y96L and Y96F mutants may arise from the ring distortions necessary for substrate binding, which could bring other ring positions into close proximity with the ferryl intermediate. The presence of the olefinic double bond in 1-phenyl-1-cyclohexylethylene must have changed the orientation of substrate binding such that although the regioselectivity for attack at C_4 was > 80% for all the Y96 mutants, the high diastereoselectivity observed with benzylcyclohexane was lost.

The results of this protein and substrate engineering work show that it is possible to broaden the substrate range of $P450_{cam}$ by mutagenesis. The observation of C–H bond activation by mutant $P450_{cam}$ proteins in molecules which also contain reactive functionalities is an important step towards utilising monooxygenases in synthesis. It appears that the $P450_{cam}$ active site will tolerate substituents of different sizes on the cyclohexane ring, and this is particularly encouraging in the context of the synthesis of polyfunctionalised cyclohexanes which are versatile synthetic intermediates.

Table 5. NADH turnover rates[a, b] of wild-type and the Y96 mutants of cytochrome $P450_{cam}$ with some phenyl derivatives

Substrate	$P450_{cam}$					
	Y96G	Y95A	Y96V	Y96L	Y96F	WT
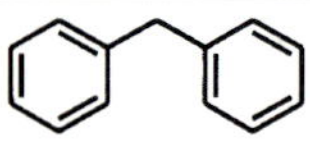	70	100	–	–	–	–
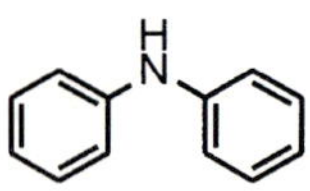	48	80	–	–	–	–
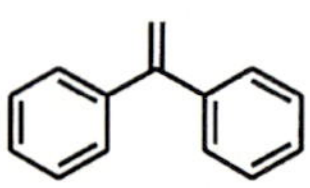	70	54	–	–	–	–
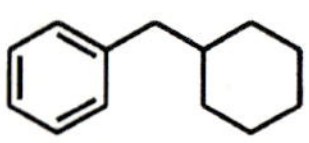	70	72	–	24	90	–
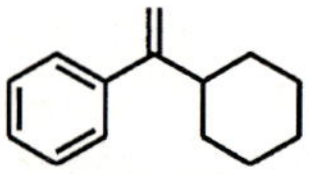	160	168	40	–	100	–

[a] The turnover rates are given as nmol NADH consumed (nmol $P450_{cam}$)$^{-1}$ (minute)$^{-1}$.
[b] –: The turnover rates are too low to be reliably calculated.

Mutant	Y96G	Y96A	Y96V	Y96L	Y96F
% of **A** in products	90	90	60	55	74

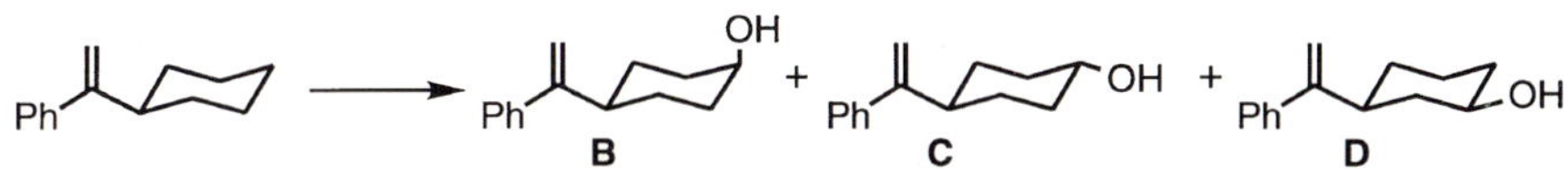

Mutant	Y96G	Y96A	Y96V	Y96L	Y96F
% **B**	47	48	41	-	39
% **C**	46	47	39	-	48
% **D**	6	5	20	-	13

Fig. 8. The distribution of products from the oxidation of benzylcyclohexane and 1-phenyl-1-cyclohexylethylene catalysed by the Y96 mutants of P450$_{cam}$

3.5 Engineering the Selectivity of Phenylcyclohexane Oxidation

The oxidation of phenylcyclohexane by wild-type and active-site mutants of P450$_{cam}$ has been investigated in an attempt to engineer the substrate hydroxylation selectivity by altering enzyme-substrate interactions [127, 128]. Phenylcyclohexane has fourteen possible hydroxylation products, allowing investigations into the effect of mutations on the selectivity of C–H bond oxidation.

Initial studies were carried out with the wild-type and Y96A and Y96F mutants of P450$_{cam}$. The three products formed were identified as *cis*-4-phenylcyclohexanol, *trans*-4-phenylcyclohexanol, and the chiral *cis*-3-phenylcyclohexanol (Table 6). The selective hydroxylation of aliphatic C–H bonds suggested that the regioselectivity was controlled by steric constraints, and that the substrate was bound with the cyclohexane ring close to the heme, with C3 and C4 nearest the iron centre. The product distributions varied significantly between the wild-type and mutant enzymes. In particular, it was noted that the product distribution of the Y96A mutant was very different from that of the wild-type and the Y96F, and it was proposed that this arose from a different substrate binding orientation in the Y96A which should have a larger substrate pocket. It was also postulated that additional mutations closer to the heme centre could be used to increase the specificity of hydroxylation, and this hypothesis was tested using the additional mutations V247A and V247L (Fig. 5).

The mutations at the 247 position were found to have no significant effect on the specificity of hydroxylation when combined with the Y96A mutation. However, when combined with Y96F, large changes in specificity were observed. This data confirmed that the phenylcyclohexane was bound in different orientations in the Y96A and Y96F mutants. The Y96F–V247L double mutant gave 83% *trans*-4-phenylcyclohexanol, whilst the Y96F–V247A gave 97% *cis*-3-phenylcyclohexanol. Of all the mutants, the Y96F–V247A double mutant also showed the highest stereoselectivity for the formation of the chiral *cis*-3-phenylcyclohexanol (42% ee). It was inferred that in the case of the Y96F-containing mutants, the phenylcyclohexane was bound in a position where it was in contact with the 247 side-chain, whereas in the Y96A-containing mutants, it was not. Thus by combining mutations at the 247 position with Y96F, it was possible to change the substrate-binding orientation and place different C–H bonds closest to the iron centre for preferential attack. These results demonstrate that it is possible to use rational protein redesign to manipulate the selectivity of hydroxylation of unactivated C–H bonds in unnatural substrates by judicious choice of active-site mutations. This clearly has wide-ranging implications in the field of protein engineering.

3.6 A "Self-Sufficient" $P450_{cam}$ Enzyme

The $P450_{cam}$ system consists of three proteins, the $P450_{cam}$ hydroxylase and the two electron-transfer proteins putidaredoxin reductase (PdR) and putidaredoxin (Pd) [26]. The $P450_{BM-3}$ from *B. megaterium* occurs naturally as a fusion protein between a heme hydroxylase domain and a reductase domain [29], and as such is described as a "self-sufficient" system. Artificial fusion proteins of mammalian P450 enzymes with their reductase domains have been expressed in yeast and *E. coli* [129, 130], and a three-component fusion protein of the mitochondrial $P450_{scc}$ system have also been constructed [131]. Similarly, Montellano and co-workers have constructed a fusion protein with all three component proteins of the $P450_{cam}$ system in a single polypeptide [132].

A number of fusion proteins were constructed in which the ordering of the genes and the length of the linker between the $P450_{cam}$ and Pd domains were varied. Catalytically active full-length fusion proteins were expressed, and the activity assays showed that the ordering of the proteins was more critical than the length of the $P450_{cam}$–Pd peptide linker. The fusion PdR–Pd–$P450_{cam}$ exhibited the highest activity, being three times as active as the $P450_{cam}$–Pd–PdR fusion. The activity of the PdR–Pd–$P450_{cam}$ fusion was only 15% of the normal reconstituted system, and detailed studies showed that the $P450_{cam}$/Pd interaction was suboptimal, resulting in slower electron-transfer. However, camphor oxidation by this self-sufficient fusion protein remained 100% coupled to NADH consumption, and this system has the advantage that it is fully active in vivo in *E. coli*. Camphor was oxidized to the normal metabolite 5-*exo*-hydroxy-

Table 6. Selectivity of the hydroxylation of phenylcyclohexane by mutants of cytochrome P450$_{cam}$

	Product distribution (%)[a]						
Products	WT	Y96A	Y96F	Y96A – V247A	Y96A – V247L	Y96F – V257A	Y96F – V247L
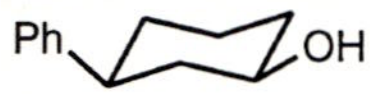	63 (36% ee)	47 (2% ee)	81 (20% ee)	45 (2% ee)	44 (2% ee)	97 (42% ee)	9 (29% ee)
	12	34	13	37	38	0	8
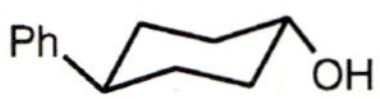	25	19	7	18	18	0	83
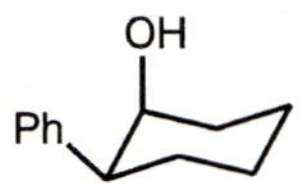	0	0	0	0	0	0	0

[a] Enantiomeric excess (ee).

camphor and some 5-oxocamphor, and it was suggested that this could arise from slow diffusion of camphor into the cytoplasm and/or of 5-*exo*-hydroxycamphor out into the growth medium.

Although the self-sufficient $P450_{cam}$ fusion protein is less active than the normal reconstituted system, this exciting work opens the door to in vivo oxidation of unnatural substrates.

4 Future Prospects

It is clear from the work discussed in this review that the detailed structural and mechanistic data on $P450_{cam}$ catalysis accumulated over the past twenty years has laid an excellent foundation for protein engineering in biotechnological applications. It is now possible to predict rationally the effect of amino acid substitutions on the coupling efficiency of the oxidation of unnatural substrates. The substrate range of the wild-type protein can be broadened by mutagenesis to include many compounds such as alkanes and substituted alicyclic compounds which are of potential importance in synthesis. The regioselectivity of the oxidation of unactivated aliphatic C–H bonds can also be engineered by judicious active site mutations. However, whilst these observations are very encouraging, the oxidation of many other classes of compounds, especially the larger biologically active molecules and pharmaceutical compounds, remains to be explored. The availability of a self-sufficient $P450_{cam}$ will allow many more in vitro and in vivo applications. Significant advances can be expected from this multi-disciplinary area of research in the near future.

5 References

1. Sligar SG, Murray RI (1986) In: Ortiz de Montellano PR (ed) Cytochrome P450: Structure, Mechanism, and Biochemistry. Plenum Press, New York, pp 429–503
2. Peterson JA, Graham-Lorence SE (1995) In: Ortiz de Montellano PR (ed) Cytochrome P450: Structure, Mechanism, and Biochemistry. 2nd edn, Plenum Press, New York, pp 151–182
3. Munro AW, Lindsay JG (1996) Mol Microbiol. 20:1115
4. Muller HG, Schunck WH, Kargel E (1991) In: Rein H, Ruckpaul K (eds) Frontiers of Biotransformation, Taylor and Francis, London; 1991, vol 4
5. Durst F (1991) In: Rein H, Ruckpaul K (eds) Frontiers of Biotransformation, Taylor and Francis, London; 1991, vol 4
6. Holtum JAM, Powles SB (ed) (1994) Herbicide Resistance in Plants: Biology and Biochemistry, Lewis Publishes, London
7. Guengerich FP (1995) In: Ortiz de Montellano PR (ed) Cytochrome P450: Structure, Mechanism, and Biochemistry. 2nd edn, Plenum Press, New York, pp 473–536
8. Ortiz de Montellano PR (ed) (1986) Cytochrome P450: Structure, Mechanism, and Biochemistry. Plenum Press, New York
9. Ortiz de Montellano PR (ed) (1995) Cytochrome P450: Structure, Mechanism, and Biochemistry. 2nd edn Plenum press, New York
10. Cardini G, Jurtshuk P (1970) J Biol Chem 245:2789
11. Asperger O, Muller R, Kleber HP (1983) Acta Biotechnol 3:319

12. Honeck H, Schunck W-H, Reige P, Muller H-G (1982) Biochem Biophys Res Commun 106:1318
13. Guengerich FP, MacDonald TL (1984) Acc Chem Res 17:9
14. Guengerich FP (1991) J Biol Chem 266:10019
15. Guengerich FP (1990) CRC Crit Rev Biochem 25:97
16. Shimada T, Martin MV, Pruess-Schwartz D, Marnett LJ, Guengerich FP (1989) Cancer Res 49:6304
17. Yun C-H, Shimada T, Guengerich FP (1992) Cancer Res 52:868
18. McManus ME, Burgess WM, Veronese ME, Huggett A, Quattrochi LC, Tukey RH (1990) Cancer Res 50:367
19. Clark BJ, Waterman MR (1991) Methods Enzymol 206:00
20. Gonzalez FJ, Kimura S, Tamura S, Gelboin HV (1991) Methods Enzymol 206:3
21. Guengerich FP, Brian WR, Sari M-A, Ross J-T (1991) Methods Enzymol 206:30
22. Porter TD, Larson JR (1991) Methods Enzymol 206:08
23. Porter TD, Coon MJ (1991) J.Biol Chem 266:3469
24. Coon MJ, Ding X, Pernecky SJ, VAZ AND (1992) FASEB J 6:69
25. Katagiri M, Ganguli BN, Gunsalus IC (1968) J Biol Chem 243:543
26. Gunsalus IC, Wagner GC (1978) Methods Enzymol 52:66
27. Mueller EJ, Loida PJ, Sligar SG (1995) Cytochrome P450: Structure, Mechanism, and Biochemistry. 2nd edn. Plenum press, New York, pp 83
28. Peterson JA, Lu JY (1991) Methods Enzymol 206:12
29. Narhi LO, Fulco AJ (1987) J Biol Chem 262:683
30. Schaiffe A, Hutchinson CR (1987) Biochemistry 26:204
31. Guengerich FP (1992) Life Sci 50:471
32. Lindberg RLP, Negishi M (1989) Nature 339:32
33. Iwasaki M, Darden TA, Parker CE, Tomer KB, Pedersen LG, Negishi M (1994) J Biol Chem 269:079
34. Yu C-A, Gunsalus IC (1974) J Biol Chem 249:4
35. Unger BP, Gunsalus IC, Sligar SG (1986) J Biol Chem 261:158
36. Peterson J-A, Lorence MC, Amarneh B (1990) J Biol Chem 265:066
37. Yasukochi T, Okada O, Hara T, Sagara Y, Sekimizu K, Horiuchi T (1994) Biochim Biophys Acta 1204:4
38. Poulos TL, Finzel BC, Gunsalus IC, Wagner GC, Kraut J (1985) J Biol Chem 260:6122
39. Poulos TL, Finzel BC, Howard AJ (1987) J Mol Biol 195:87
40. Raag R, Poulos TL (1991) Biochemistry 30:674
41. Poulos TL, Howard AJ (1987) Biochemistry 26:165
42. Raag R, Li H, Jones BC, Poulos TL (1993) Biochemistry 32:571
43. Poulos TL, Finzel BC, Howard AJ (1986) Biochemistry 25:314
44. Sligar SG, Gunsalus IC (1976) Proc Natl Acad Sci USA 73:078
45. Sligar SG (1976) Biochemistry 15:399
46. Fisher MT, Sligar SG (1985) J Am Chem Soc 107:5018
47. Harris D, Loew GH (1993) J Am Chem Soc 115:775
48. Hintz MJ, Peterson JA (1981) J Biol Chem 256:721
49. Hintz MJ, Mock DM, Peterson LL, Tuttle K, Peterson JA (1982) J Biol Chem 257:4324
50. Stayton PS, Fisher MT, Sligar SG (1989) J Biol Chem 263:13544
51. Stayton PS, Poulos TL, Sligar SG (1989) Biochemistry 28:8201
52. Stayton PS, Sligar SG (1990) Biochemistry 29:7381
53. Geren L, Tuls J, O'Brien P, Millett F, Peterson JA (1986) J Biol Chem 261:15491
54. Koga H, Sagara Y, Yaoi T, Tsujimura M, Nakamura K, Sekimizu K, Makino R, Shimada H, Ishimura Y, Yura K, Go M, Ikeguchi M, Horiuchi T (1993) FEBS Lett 331:109
55. Davis MD, Sligar SG (1994) Biochemistry 31:11383
56. Kazlauskaite J, Westlake ACG, Wong L-L, Hill HAO (1996) J Chem Soc, Chem Commun 2189
57. Sligar SG, Lipscomb JD, Debrunner PG, Gunsalus IC (1974) Biochem Biophys Res Commun 61:290
58. Peterson JA, Ishimura Y, Griffin BW (1972) Arch Biochem Biophys 149:197

59. Groves JT, McClusky GA, White RE, Coon MJ (1978) Biochem Biophys Res Commun 81:154
60. Groves JT, McClusky GA (1976) J Am Chem Soc 98:859
61. Atkins WM, Sligar SG (1988) Biochemistry 27:1610
62. Kadkhodayan S, Coulter ED, Maryniak DM, Bryson TA, Dawson JH (1995) J Biol Chem 270:28042
63. Gelb MH, Heimbrook DC, Malkonen P, Sligar SG (1982) Biochemistry 21:370
64. White PW (1990) Bioorg Chem 18:440
65. Gunter MJ, Turner P (1991) Coord Chem Rev 108:115
66. Mansuy D, Battioni P (1989) Alkane Functionalization by cytochromes P450 and by model systems using O_2 or H_2O_2. In: Hill CL (ed) Activation and Functionalization of Alkanes. Wiley, New York, p 195
67. Pratt J, Ridd TI, King LJ (1995) J Chem Soc, Chem Commun 2297
68. Lee-Robichaud P, Shyadehi AZ, Wright JN, Akhtar ME, Akhtar M (1995) Biochemistry 34:10104
69. Roberts ES, Vaz AND, Coon MJ (1991) Proc Natl Acad USA 88:8963
70. Gould P, Gelb MH, Sligar SG (1981) J Biol Chem 256:6686
71. Gelb MH, Heimbrook DC, Mälkönen, P, Sligar SG (1982) Biochemistry 21:370
72. Filipovic D, Paulsen MD, Loida PJ, Sligar SG, Ornstein RL (1992) Biochem Biophys Res Commun 189:488
73. Ortiz de Montellano PR, Fruetel JA, Collins JR, Camper DL, Loew GH (1991) J Am Chem Soc 113:3195
74. Atkins WM, Sligar SG (1987) J Am Chem Soc 109:3754
75. Gorsky LD, Koop DR, Coon MJ (1984) J Biol Chem 259:6812
76. Kuthan H, Ullrich V (1982) Eur J Biochem 126:583
77. Loida PJ, Sligar SG (1993) Biochemistry 32:11530
78. Haseman CA, Ravichandran KG, Peterson JA, Deisenhofer J (1994) J Mol Biol 236:1169
79. Cupp-Vickery J, Poulos TL (1995) Nat Struct Biol 2:144
80. Ravichandran KG, Boddupalli SS, Hasemann CA, Peterson JA, Deisenhofer J (1993) Science 261:731
81. Poulos TL, Cupp-Vickery J, Li H (1995) In: Ortiz de Montellano PR (ed) Cytochrome P450: Structure, Mechanism, and Biochemistry. Plenum Press, New York, pp 125
82. Hasemann CA, Kurumbail RG, Boddupalli SS, Peterson JA, Deisenhofer J (1995) Structure 3:41
83. Nicholls A, Sharp K, Honig B (1991) Prot Struct Funct Gen 11:281
84. Kraulis PJ (1991) J Appl Cryst 24:946
85. Bacon DJ, Anderson WF (1988) J Molec Graphics 6:219
86. Merritt EA, Murphy MEP (1994) Acta Cryst D50:869
87. Griffin BW, Peterson JA (1972) Biochemistry 11:4740
88. Raag R, Poulos TL (1989) Biochemistry 28:7586
89. Laughton CA, Neidle S, Zvelebil MJJM, Sternberg MJE (1990) Biochem Biophys Res Commun 171:1160
90. Koymans L, Vermeulen N, Baarslag A, Denkelder G (1993) J Comput Aided Mol Des 281
91. Boscott PE, Grant GH (1994) J Mol Graph 12:185
92. Lewis DFV, Ioannides C, Parke DV (1994), Toxicol. Letts 71:235
93. Degroot MJ, Vermeulen N, Kramer JD, Vanacker F, Denkelder G (1996) Chem Res Toxicol 1079
94. Graham-Lorence S, Amarneh B, White RE, Peterson JA, Simpson ER (1995) Protein Science, vol 1065
95. Koymans LMH, Moereels H, Vanden-Bossche H (1995) J Steroid Biochem Mol Biol 53: 191
96. Szklarz GD, He YA, Halpert JR (1995) Biochemistry 34:14312
97. Braatz JA, Bass MB, Ornstein RL (1994) J. Comput. Aided Mol Des 8:607
98. Szklarz GD, Ornstein RL, Halpert JR, (1994) J Biomol Struct Dyn 12:61
99. Lin D, Zhang LH, Chiao E, Miller WL (1994) Mol Endocrinol 8:392
100. De Voss JJ, Ortiz de Montellano PR (1995) J Am Chem Soc 117:4185

101. Sligar SG, Filipovic D, Stayton PS (1991) Methods Enzymol 206:31
102. Paulsen MD, Bass MB, Ornstein RL (1991) J Biomol Struct Dyn 9:187
103. Swanson BA, Dutton DR, Lunetta JM, Yang CS, Ortiz de Montellano PR (1991) J Biol Chem 266:19258
104. Tuck SF, Graham-Lorence S, Peterson JA, Ortiz de Montellano PR (1993) J Biol Chem 268:269 and references therein
105. Atkins WM, Sligar SG (1988) J Biol Chem 263:18842
106. Fisher MT, Sligar SG (1987) Biochemistry 26:4797
107. Atkins WM, Sligar SG (1989) J Am Chem Soc 111:2716
108. Martinis SA, Atkins WM, Stayton PS, Sligar SG (1989) J Am Chem Soc 111:9252
109. Imai M, Shimada H, Watanabe Y, Matsushima-Hibaya Y, Makino R, Koga H, Horiuchi T, Ishimura Y (1989) Proc Natl Acad Sci USA 86:7823
110. Kimata Y, Shimada T, Ishimura Y (1995) Biochem. Biophys Res Commun 208:96
111. Raag R, Martinis SA, Sligar SG, Poulos TL (1991) Biochemistry 30:11420
112. Aikens J, Sligar SG (1994) J Am Chem Soc 116:1143
113. Gerber NC, Sligar SG (1992) J Am Chem Soc 114:8742
114. Gerber NC, Sligar SG (1994) J Biol Chem 269:4260
115. Shimada H, Makino R, Unno M, Horiuchi T, Ishimura Y (1993) In: Lechner MC (ed) Cytochrome P450 Biochemistry, Biophysics and Molecular Biology, John Libbey, pp 299
116. Loida PJ, Sligar SG (1993) Protein Eng 2:207
117. Fruetel JA, Collins JR, Camper DL, Loew GH, Ortiz de Montellano PR (1992) J Am Chem Soc 114:6987
118. Watanabe Y, Ishimura Y (1989) J Am Chem Soc 111:410
119. Jones JP, Trager WF, Carlson TJ (1993) J Am Chem Soc 115:381
120. Grayson DA, Tewari YB, Mayhew MP, Vilker VL, Goldberg RN (1996) Arch Biochem Biophys 332:239
121. Castro CE, Wade RS, Belser NO (1985) Biochemistry 24:204
122. Li S, Wackett LP (1993) Biochemistry 32:9355
123. Lefever MR, Wackett LP (1994) Biochem Biophys Res Commun 201:373
124. Stevenson JA, Westlake ACG, Whittock C, Wong L-L (1996) J Am Chem Soc, 118:12846
125. Fowler SM, England PA, Westlake ACG, Rouch DA, Nickerson DP, Blunt C, Braybrook D, West S, Wong L-L, Flitsch SL (1994) J Chem Soc, Chem Commun 2761
126. Bell SG, Rouch DA, Wong LL, J Mol Catal, in press
127. England PA, Rouch DA, Westlake ACG, Bell SG, Nickerson DP, Webberley M, Flitsch SL, Wong L-L (1996), J Chem Soc, Chem Commun 357
128. Jones NE, England PA, Rouch DA, Wong L-L (1996) J Chem Soc Chem Commun 2413
129. Murakami H, Yabusaki Y, Sakaki T, Shibata M, Ohkawa H (1987) DNA Cell Biol 6:189
130. Shet MS, Faulkner KM, Holmans PL, Fisher CW, Estabrook RW (1995) Arch Biochem Biophys 318:314
131. Harikrishner JA, Black SM, Szklarz GD, Miller WL (1993) DNA Cell Biol 12:371
132. Sibbesen O, De Voss JJ, Ortiz de Montellano PR (1996) J Biol Chem 271:22462

Author Index Volumes 1–88

Printing: Saladruck, Berlin
Binding: Buchbinderei Lüderitz & Bauer, Berlin